连接：共享未来
thisCONNECT
SHARING A FUTURE
2017 上海城市空间

上海城市空间艺术季展览画册编委会 编

来的公共空间
NECTION
RE PUBLIC SPACE
间 艺 术 季 主 展 览

同济大学出版社
TONGJI UNIVERSITY PRESS

2017上海城市空间艺术季主展览
2017 SHANGHAI URBAN SPACE ART SEASON MAIN EXHIBITION

主办
上海市城市雕塑委员会

承办
上海市规划和国土资源管理局
上海市文化广播影视管理局
上海市浦东新区人民政府

协办
上海地产（集团）有限公司
上海东岸投资（集团）有限公司

Host
Shanghai Urban
Sculptures Committee

Organizer
Shanghai Municipal
Bureau of Planning and
Land Resources
Shanghai Municipal
Administration of Culture,
Radio, Film & TV
People's Government of
Pudong New District,
Shanghai

Co-organizer
Shanghai Land (Group)
CO., LTD
Shanghai East Bund
Investment (Group)
CO., LTD

编委会成员
EDITORIAL BOARD

主编	CHIEF EDITORS
徐毅松	XU Yisong
于秀芬	YU Xiufen
杭迎伟	HANG Yingwei

副主编	DEPUTY CHIEF EDITORS
王训国	WANG Xunguo
吴孝明	WU Xiaoming
诸迪	ZHU Di
张玉鑫	ZHANG Yuxin

执行主编	EXECUTIVE CHIEF EDITORS
方振宁	FANG Zhenning
李翔宁	LI Xiangning
赵宝静	ZHAO Baojing

编委 EDITORIAL BOARD MEMBERS
按姓氏笔画排序 *SORTED BY SURNAME STROKES*

支文军	ZHI Wenjun
王明颖	WANG Mingying
冯路	FENG Lu
关也彤	GUAN Yetong
沈捷	SHEN Jie
陈毅国	CHEN Yiguo
林磊	LIN Lei
侯斌超	HOU Binchao
郭晓彦	GUO Xiaoyan
徐孙庆	XU Sunqing
黄蕴菁	HUANG Yunjing
蒋卫中	JIANG Weizhong
戴春	DAI Chun

学术委员会

为确保空间艺术季的专业性、学术性和国际性，特成立空间艺术季学术委员会，为空间艺术季各项核心工作提供专业指导和评审意见。委员会体现国际性和专业性，委员均为国际和国内专业造诣高、工作经验丰富且有影响力的专家，并保证国际专家占一定的比例。委员会体现了多专业的跨界合作，充分体现空间艺术季是专业性、公众性相结合的城市大型公共活动。

职责

学术委员会职责包括召开各阶段学委会会议，审议有关工作事项，进行策展人和策展方案评审等；活动期间参与论坛和研讨会，进行主旨演讲和学术交流等；出席空间艺术季期间各重要节点活动等。

委员

学术委员会委员由包括城市规划、建筑、景观、公共艺术、传播、评论、策展、社会学、媒体等领域的专家组成，2017上海城市空间艺术季学术委员会委员共31人。

ACADEMIC COMMITTEE

To ensure SUSAS as a professional, academic and international event, the academic committee is established to offer professional guides and evaluative opinions on various core tasks of the art season. Committee members, domestic and abroad, feature expertise, experiences and influence, and a considerable number of them are internationally renowned. The committee is comprised of members from multiple disciplines, fully demonstrating SUSAS as a major urban public event that is both professional and open.

RESPONSIBILITY

Responsibilities of the academic committee include: organizing academic committee sessions at each stage where progress, curators and plans are discussed and evaluated; attending forums and seminars during the art season, giving keynote speeches and participating in academic dialogues; attending various key activities during the art season, et cetera.

MEMBERS

Members of the academic committee cover 31 professionals in urban planning, architecture, landscaping, public art, communication studies, art criticism, curation, sociology and the press. The list is subject to change according to each year's requirements.

学术委员会主任	PRESIDENTS
郑时龄	ZHENG Shiling
吴为山	WU Weishan
学术委员会委员	COMMISSIONER
按姓氏笔画排序	SORTED BY SURNAME STROKES
于海	YU Hai
王林	WANG Lin
王才强（新）	Heng Chye Kiang (Singapore)
王建国	WANG Jianguo
北川富朗（日）	Fram Kitagawa (Japan)
朱子瑜	ZHU Ziyu
伍江	WU Jiang
孙玮	SUN Wei
芭芭拉·菲舍尔（加）	Barbara Fischer (Canada)
李磊	LI Lei
李向阳	LI Xiangyang
李振宇	LI Zhenyu
杨劲松	YANG Jinsong
杨奇瑞	YANG Qirui
汪大伟	WANG Dawei
沈迪	SHEN Di
张杰	ZHANG Jie
张永和	Yung Ho Chang
张宇星	ZHANG Yuxing
郑佳矢	ZHENG Jiashi
郑培光	ZHENG Peiguang
赵宝静	ZHAO Baojing
俞孔坚	YU Kongjian
莫森·穆斯塔法维（美）	Mohsen Mostafavi (USA)
殷小烽	YIN Xiaofeng
曹嘉明	CAO Jiaming
鲁晓波	LU Xiaobo
曾成钢	ZENG Chenggang
戴维·汤普森（美）	David Thompson (USA)

目 录

026 序一 / 徐毅松 上海市规划和国土资源管理局局长
028 序二 / 吴孝明 上海市文化广播影视管理局艺术总监
030 序三 / 张玉鑫 浦东新区副区长

032 郑时龄 / 连接城市空间，以心点燃人们的心

036 斯坦法诺·博埃里 / 从公共艺术空间连接未来城市
042 李翔宁 / 超越展览的展览：作为城市的介入方式
046 方振宁 / 场所决定形态：SUSAS 2017 艺术介入计划

056 柳亦春 / 时间与地点的再定义：民生码头八万吨筒仓建筑的临时性改造与再利用
060 主题 thisCONNECTION 阐释：连接的多义性

068 P/ 特展
132 T/ 公共空间形态
182 H/ 社会文化多样
244 I/ 基础设施连接
272 S/ 上海都市范本

312 展览项目索引
316 SUSAS 学院
325 策展团队
326 花絮
336 图片版权与摄影者索引

337 致谢

P / 特展

- 070 P1 建造当代的文化图景
- 076 P2 巴塞罗那：城市群大都会
- 080 P3 拉斯维加斯工作室：来自罗伯特·文丘里和丹尼斯·斯科特·布朗档案馆的影像集
- 084 P4 液态历史：泰晤士河的想象与现实
- 088 P5 当代中国的多元建筑实践
- 096 P6 与水共生：世界优秀水岸空间案例展
- 106 P7 社会图景：来自城市内部的影像学
- 110 P8 木构与智构
- 114 P9 漫步环翠堂园景
- 118 P10 马列维奇视觉年表
- 124 P11 万象
- 128 P12 鼓浪屿历史国际社区——共享遗产保护之路

T / 公共空间形态

- 134 T1 林中之境
- 142 T2 2017年欧盟当代建筑奖——密斯·凡·德·罗奖和2016年"Fear of Columns"竞赛展览
- 148 T3 全球建筑实践罗盘：一种新兴建筑的分类学
- 150 T4 哥伦布之谜系列雕塑展
- 152 T5 凝聚
- 156 T6 混乱中迷失
- 162 T7 风卷
- 164 T8 中国文人写意雕塑园（9件）
- 170 T9 内省腔
- 174 T10 设计长椅
- 176 T11 风律
- 178 T12 仓声·品
- 180 T13 2340洞

H / 社会文化多样

- 184 H1 回音：建筑与社会
- 190 H2 南京长江大桥记忆计划
- 194 H3 濑户内国际艺术祭
- 198 H4 连接：空间移动
- 214 H5 空间的边界
- 218 H5 你是如此温柔、幻想速度
- 222 H5 寻找马列维奇
- 226 H5 天路
- 230 H5 蝴蝶夫人
- 234 H5 小库：人工智能时代的未来都市
- 238 H6 内心（影像）

I / 基础设施连接

- 246 I1 桥
- 258 I2 物联生产
- 262 I3 体验HyperloopTT
- 266 I4 数字建造：数字金属
- 268 I4 数字建造：超薄纸板大跨建构
- 270 I4 数字建造：机器人木构

S / 上海都市范本

- 276 S1 上海城事
- 281 S2 城市微空间复兴计划
- 286 S3 上海空间研究
- 289 S4 新村研究
- 293 S5 两岸贯通
- 300 S6 文化点亮城市
- 308 S7 更新上海

CONTENTS

026 PREFACE I / XU Yisong *Director of Shanghai municipal bureau of planning and land resources*
028 PREFACE II / WU Xiaoming *Art director of Shanghai municipal bureau of culture, radio, film & TV*
030 PREFACE III / ZHANG Yuxin *Deputy district mayor of the People's Government of Pudong New District*

032 ZHENG Shiling / A CONNECTION OF URBAN SPACE, A TORCH TO LIGHTEN

036 Stefano Boeri / "thisCONNECTION": FROM PUBLIC ART SPACE TO A FUTURE CITY
042 LI Xiangning / EXHIBITION BEYOND EXHIBITION: A WAY TO INTERVENE CITIES
046 FANG Zhenning / PLACE DICTATES FORM: ART INTERVENTION IN SUSAS 2017

056 LIU Yichun / REDEFINING TIME AND PLACE: TEMPORAL REPURPOSING AND REUSE OF THE 80,000-TON SILO AT MINSHENG PORT
060 AN INTERPRETATION OF "thisCONNECTION": POLYSEMY OF CONNECTION

068 P / SPECIAL PAVILION
132 T / TOPOLOGY
182 H / HETEROGENEITY
244 I / INFRASTRUCTURE
272 S / SHANGHAI SAMPLE

314 EXHIBITION ITEM INDEX
316 SUSAS COLLEGE
325 CURATORS
326 TIDBITS
336 COPY RIGHTS AND PHOTOGRAPHY CREDITS

337 ACKNOWLEDGEMENTS

P / SPECIAL PAVILION

- 070 P1 Constructing a Contemporary Cultural Landscape
- 076 P2 Barcelona, Metropolis of Cities
- 080 P3 Las Vegas Studio: Images from the Archives of Robert Venturi and Denise Scott Brown
- 084 P4 Liquid Histories: the Thames, between the Real and Imaginary
- 088 P5 Diverse Practices in Contemporary Chinese Architecture
- 096 P6 Living with Water: World Outstanding Waterfront Space Case Exhibition
- 106 P7 Social View: Iconography from City
- 110 P8 Wooden Structures and Smart Structures
- 114 P9 Stroll in Huancui Tang
- 118 P10 Malevich Visual Chronology
- 124 P11 Social View: Wanxiang
- 128 P12 A Sharing Conservation Approach: Kulangsu, a Historic International Settlement

T / TOPOLOGY

- 134 T1 Radura
- 142 T2 2017 The EU Mies Award and 2016 "Fear of Columns"
- 148 T3 Architecture's "Political Compass": A Taxonomy of Emerging Architecture
- 150 T4 The Mysteries of Columbus Cristobal
- 152 T5 Cohesion
- 156 T6 LOST IN A SHUFFLE
- 162 T7 Rolling Wind
- 164 T8 Chinese Freestyle Scholars' Sculpture Park
- 170 T9 Introspective Cavity (exterior)
- 174 T10 Design Bench
- 176 T11 Rhythm of Wind
- 178 T12 Cang Sheng · Pin
- 180 T13 2340 Holes

H / HETEROGENEITY

- 184 H1 The Echo from Society: Architecture and Contemporary Challenges out of Established Agendas
- 190 H2 Memory Project of the Nanjing Yangtze River Bridge
- 194 H3 Setouchi Triennale
- 198 H4 Connection: Space Movement
- 214 H5 The Frontier of Space
- 218 H5 You are so Tender, Fantasy Speed
- 222 H5 Malevich Finders
- 226 H5 Cosmos
- 230 H5 Madame Butterfly
- 234 H5 Future Metropolitan of AI Era
- 238 H6 Inner Space

I / INFRASTRUCTURE

- 246 I1 Bridge
- 258 I2 Interconnected in Production
- 262 I3 Hyperloop TT Experience
- 266 I4 Digital Fabrication: Digital Metal
- 268 I4 Digital Fabrication: Shells with Thin Sheet Materials
- 270 I4 Digital Fabrication: Robotic Timber Construction

S / SHANGHAI SAMPLE

- 276 S1 Shanghai Practice
- 281 S2 Urban Humble-Space Regeneration
- 286 S3 Shanghai Urban-Space Research
- 289 S4 Study of New Villages
- 293 S5 Waterfront Connection
- 300 S6 Culture Englightens the City
- 308 S7 Renewing Shanghai

序一 PREFACE I

徐毅松 上海市规划和国土资源管理局局长

XU Yisong director of Shanghai municipal bureau of planning and land resources

2017 上海城市空间艺术季在美丽的黄浦江畔的民生码头顺利开幕。本届艺术季由上海市城市雕塑委员会主办，上海市规划和国土资源管理局、上海市文化广播影视管理局及当届主展览所在地浦东新区人民政府共同承办。

民生码头 8 万吨筒仓曾是亚洲最大的粮仓，随着工业时代的繁华褪去，码头已逐渐淡出人们的视野。2017 上海城市空间艺术季于 2017 年 10 月 15 日至 2018 年 1 月 15 日在这里对公众免费开放，展期三个月，这个沉寂多年的工业遗址因观众的到来而重焕生机与活力。

本届空间艺术季的筹办过程凝聚了全球各界精英的智慧和心血。远道而来的策展人、参展艺术家、建筑师、规划师、各行各业的学者专家，以及一直关注空间艺术季的各方媒体，纷纷为空间艺术季出谋划策，为上海城市发展贡献自己的力量，使空间艺术季真正成为众创共享的市民文化盛宴。

芬兰建筑师伊里尔·沙里宁（Eliel Saarinen）曾说："让我看看你的城市，我就能说出这个城市的居民在文化上追求什么。"为实现 2035 年建设"卓越的全球城市"这一目标愿景，上海不仅要打造具有全球影响力的城市功能，还要塑造平等、包容、更富人文关怀、更具自身鲜明特质的城市文化。上海城市空间艺术季作为一个公众性的平台和窗口，将向世界展示上海独具特色的人文魅力和精彩纷呈的城市空间活力。

空间艺术季是一项将展览与实践相结合、"双年展制"的城市空间艺术展示活动。它以"文化兴市，艺术建城"为理念，以"城市公共空间品质提升"为长期固定主题，致力于通过打造具有"国际性、公众性、实践性"的城市空间艺术品牌活动，从而改善城市空间品质、提升城市魅力，实现"城市艺术化，艺术生活化"的目标。

2017 上海空间艺术季由主展览、实践案例展、联合展和 SUSAS 学院共同构成。主展览以"连接 thisCONNECTION：共享未来的公共空间"为主题，含四大主题展、12 个特展、200 多个展项作品，约 200 位来自全球 17 个国家和地区的规划师、建筑师、策展人、艺术家等共同参与。与上届相比，本届空间艺术季主展览希望在三个方面有所突破。一是聚焦城市空间服务于民的理念，聚焦于城市空间的连接，如大众瞩目的黄浦江滨江贯通工程、城市微更新等。二是注重以城市空间作展场，坚持展览与实践相结合，实现展示活动每举办一届，城市公共空间就美化一片，文化热点就传播一次，国内外大师作品就沉淀一批的目的。例如将主展场设置在民生码头 8 万吨筒仓及周边开放空间，让筒仓改建本身成为展览中最大一件展品，成为滨江新的艺术核心。三是积极搭建公众参与平台，促进公共教育，注重社会效应，既坚持学术性、思想性和前瞻性，通过国际征集，确定米兰理工大学教授、著名建筑师斯坦法诺·博埃里、同济大学建筑与城规学院副院长李翔宁、独立策展人方振宁作为主策展团队；同时又注重在市民中增加影响力，在不到两个月的时间内有超过 10 万的参观者，也体现了公众对于展览的关注，对于规划设计如何改变生活方式、提升生活品质的关注。

除了主展览外，全市范围内还有 8 个实践案例展和 6 个联合展，其中实践案例展包括历史风貌的保护和重塑、滨水等城市公共空间复兴、社区空间营造和更新三大类型，都是市民家门口触手可及的空间艺术和最深切感受城市温度的触手。此外，展期内还举行了 100 多场 SUSAS 学院和其他公众活动。

亚里士多德曾说："人们为了活着，聚集于城市；人们为了活得更好，居留于城市。"城市是文化的载体，文化是城市的灵魂。相比城市物质建设的速度，富有魅力的人文环境塑造也许需要经历更加漫长的过程。希望通过本届城市空间艺术季的举办，继承发扬"城市，让生活更美好"的世博精神，推动上海建设成为"更富魅力的幸福人文之城"！

Shanghai Urban Space Art Season 2017 (SUSAS 2017) is opened at Minsheng Port on the banks of the Huangpu River. It is hosted by Shanghai Urban Sculptures Committee, and organized by Shanghai Municipal Bureau of Planning and Land Resources, Shanghai Municipal Administration of Culture, Radio, Film & TV, and People's Government of Pudong New District, where the events will take place.

The 80,000-ton Silo at Minsheng Port used to be the largest one used for bulk grain in Asia, and has gradually lost its glamor with Shanghai's industrial past. SUSAS 2017 lasts three months between October 15, 2017 and January 15, 2018, and is freely accessible to all citizens, reinvigorating this long-forgotten industrial heritage.

This event crystalizes dite efforts and insights from countries and sectors, including curators, artists, architects, planners, researchers of various disciplines and the press. Thanks to their plans and contributions to the city, SUSAS has become a public cultural event created and enjoyed by all.

Eliel Saarinen, a Finnish architect, once said, "Take me to your city, and I can tell the cultural desires of those who live there." To become an "outstanding global city" by 2035, Shanghai needs as much an egalitarian, tolerant, humane and salient culture unique to the city as metropolitan functions of global influence. As a public platform, SUSAS aims to present the city's cultural features and vibrant urban spaces with flair.

SUSAS is a biennale that combines exhibitions and site projects, to demonstrate urban space art. Its core idea is "Culture Enriches City, Art Enlightens Space", with its long-term theme "Improve the Quality of Urban Public Spaces". Dedicated to creating a permanent brand of urban space art, it seeks to become "international, public and practical" for the sake of better urban spaces and a more charming city, with the aim of becoming a "city of art" and "art of life".

SUSAS 2017 comprises the Main Exhibition, Site Projects, Joint Exhibitions and SUSAS College. Guided by the theme "thisCONNECTION: Sharing a Future Public Space", the Main Exhibition includes four theme-related sub-exhibitions, twelve special pavilions and more than two hundred items, involving about 200 planners, architects, curators and artists from 17 countries and regions. It aims to focus on three aspects under-represented in the last SUSAS. The first are innovative ideas that urban spaces may serve the citizenry, particularly in how to connect various spaces. Prominent examples are the Waterfront Connection Project of the Huangpu River and urban micro-regenerations. The second is using urban spaces as natural exhibition halls, in accordance with the principle that exhibitions and practical life should be integrated. Each SUSAS has the basic task to beautify a series of public spaces, create a cultural hotspot and commission long-standing masterpieces by domestic and foreign artists. For example, the 80,000-ton Silo, the largest in this exhibition, and its surrounding public spaces will be the new art nexus of the Huangpu River waterfront. The third is to build a public participatory platform as a faculty of general education and social well-being, while sticking to its natural inclination as an academic, insightful and forward-looking event. The main curators, the result of international competition, includes Stefano Boeri (professor of Politecnico di Milano and architect), LI Xiangning (Vice Dean of College of Architecture and Urban Planning, Tongji University) and Fang Zhenning (independent curator). The fact that it received more than 100,000 visitors in less than two months has shown the citizens' interest in the exhibition itself, and more generally, their concern about how urban design may alter and improve their life.

Aside from the main exhibition, eight site projects and six joint exhibitions are held throughout the city. These site projects are categorized into "Protection and Reshaping of Historical Heritages", "Reinvigoration of Waterfront Public Spaces and Others", and "Creation and Re-creation of Community Spaces", all of which are situated in residential areas so that they provide the public with an up-close experience of these works of space art. various public events and those of SUSAS College are held during this three-months period, with more than one hundred events offered.

As Aristotle once said, "People gather in cities first and foremost for security, then for economic opportunity, and then stay for the good life". Cities embody culture, while culture makes cities soulful. Compared to physical constructions, a charming, cultural environment may need more time to emerge. I hope that SUSAS 2017 will inherit and enhance the spirit of the Shanghai Expo, "Better City, Better Life", and contribute to Shanghai as a "city of cultural attractiveness, happiness and humanity".

序二
PREFACE II

吴孝明 上海市文化广播影视管理局 艺术总监

WU Xiaoming Art director of Shanghai municipal bureau of culture, radio, film & TV Artistic director

上海城市空间艺术季以"文化兴市，艺术建城"为理念，打造具有"国际性、公众性、实践性"的城市空间艺术品牌活动，从而实现"城市艺术化，艺术生活化"的目标。正如市委书记韩正同志所强调的，要把文化看作上海这座城市的人民福祉、活力魅力所在、竞争力的本质体现。发展上海的文化应该多种举措并举，营造更浓厚的文化氛围。

近年来，为进一步提升城市空间文化品质，激活城市公共文化空间，上海市文广局围绕上海建设"四个中心"、国际文化大都市的目标，大力实施文化与城市规划、交通、商业等融合发展的战略，积极营造与上海城市更新、城市发展相匹配的城市文化建设。2016 年发布第二个《营造上海城市文化氛围三年行动计划》以来，市文广局利用城市主要广场、绿地等公共空间，着力打造了城市广场音乐会；利用地铁站厅、通道、车厢等公共空间，引入雕塑、绘画、摄影、书法展览和街头艺人表演等公共艺术形式，打造精品地铁文化；利用南京路、徐家汇、虹桥、五角场、陆家嘴等商业集聚区域，引入美术、非遗、演艺等各类展览、展示、展演，打造文商互动的公共空间艺术，初步形成了社会主体参与广泛、文化活动丰富多彩、艺术空间随处可及、广大市民分享便利的上海公共文化氛围。

本届艺术季主展览以"连接 thisCONNECTION：共享未来的公共空间"为主题，包含四大主题展、12 个特展，有 200 多位设计师、艺术家、建筑师等国内外参展人共同参与。除了主展览外，全市范围内还有 8 个实践案例展和 6 个联合展。8 个实践案例展选取与百姓生活密切相关的公共空间，从建设发展实际出发，通过策展与市民文化活动等多种方式，展示城市空间公共艺术改造方案及实施效果，吸引更多的市民关注城市空间的发展和变化，共同提升空间活力和品质。

在空间艺术季活动期间，市文广局继续落实"营造上海城市文化氛围三年行动计划"，开启"艺术 + 商业"联动发展的创新模式。引进高雅艺术项目入驻全市商圈、购物中心、特色商业街区等业态，推动商业设施由传统模式向体验经济转型，同时打造城市公共空间的文化厚度及人文情怀，提升市民文化素养。用多元社会主体的创意，去参与创造城市空间、激发全社会共同塑造、共同美化城市空间，为党的十九大胜利召开营造了良好的氛围。

我们希望广大市民用眼睛、文字、镜头，去发现、传播上海城市空间的艺术美、城市居民的生活美，共同参与到上海城市空间艺术的成果分享、问题思考和未来谋划中来。

Guided by the idea "Culture Enriches a City, Art Enlightens Space", Shanghai Urban Space Art Season (SUSAS) is dedicated to creating a permanent brand of urban space art that is "international, public and practical" with the aim of becoming a "city of art" and "art of life". As stressed by Zheng Han, the Secretary Municipal Committee of the CPC of Shanghai, culture should be considered as an essential reflection of the city's vigor, charm and competition, and a variety of measures should be implemented to develop Shanghai's cultural enterprise and create a robust cultural atmosphere.

In recent years, to improve the culture of urban spaces and activate urban public cultural spaces, the Shanghai Municipal Bureau of Culture, Radio, Film & TV (SMBCRFT) has implemented a series of measures aiming to build "four centers" in Shanghai and transform it into a "international cultural metropolitan" area, including a strategy to incorporate culture with city planning, transportation and business, and cultural projects in accordance with Shanghai's regeneration and development. Since the second "Three-Year Action Plan to Build Cultural Atmosphere in Shanghai" was published in 2016, SMBCRFT has preliminarily created a public cultural atmosphere featuring comprehensive participation, multitudinous events, pervasive art spaces and citizen availability through a series of projects, including plaza concerts in squares and parks; metro culture, that is, sculptures, paintings, photography, calligraphy and street shows in stations, pedestrian walkways and carriages; public art spaces in core business areas like Nanjing Road, Xujiahui, Hongqiao, Wujiaochang and Lujiazui, where a variety of exhibitions, shows and performances are held, showcasing fine art, intangible cultural heritage, concerts and popular performance.

Guided by the theme "thisCONNECTION: Sharing a Future Public Space", the Main Exhibition of SUSAS 2017 comprises four themed sub-exhibitions and twelve special pavilions, which have crystalized the efforts of more than 200 designers, artists and architects from around the world and China. Apart from the Main Exhibition, eight projects and six joint exhibitions took place throughout the city. The site projects are all public spaces immersed in daily life. Based on the actual construction and thoughtful curation as well as public cultural activities, they're intended to demonstrate plans and effects of public art renovation, attracting more citizens to the development and transformation of urban spaces and participate in their improvement.

During the Art Season, SMBCRFT will continue to implement the "Three-Year Action Plan to Build Up Cultural Atmosphere in Shanghai", and launch an innovative pattern to interlink art and business. Many fine art projects will be stationed in business districts, shopping malls and street markets around the city to transform them from conventional retail areas to an experienced economy space, as well as a measure to promote cultural and humane values of public spaces and enhance citizens' cultural taste. With the upcoming 19th CPC National Congress, these measures are calling forth ideas and efforts from various parties of our society for creating, shaping and beautifying urban spaces.

We hope that the public can spot and help spread the aesthetics of Shanghai's urban spaces as well as the fascination of its citizenry through their eyes, words and lens. In doing this, they share, reflect and plan Shanghai's urban space art.

序三
PREFACE III

张玉鑫 浦东新区副区长

ZHANG Yuxin Deputy district mayor of the People's Government of Pudong New District

金秋十月，2017上海城市空间艺术季顺利开幕。走进主展览，我们为之震撼。民生码头，曾经的散粮、散糖装卸码头；8万吨筒仓，曾经的亚洲最大粮仓，早已褪去了工业时代的繁华。如今，沉寂多年的民生码头随着黄浦江两岸公共空间的贯通而焕发新生，这座气势磅礴的8万吨筒仓成了2017上海城市空间艺术季的主展场。

为此，浦东新区上海东岸投资（集团）对8万吨筒仓进行了保护性改造，尽最大努力保留筒仓的原本风貌，将底层和最顶层的空间整合为同时使用的展览空间，激发了筒仓的艺术活力。我们在筒仓壁外挂了一组自动扶梯，将人流直接引至顶层展厅，扶梯以轻盈简约的形式融入筒仓建筑，既满足交通疏散需求，又能让人们在参展的同时欣赏到北侧黄浦江以及整个民生码头的壮丽景观，塑造了沿江景观空间。与此同时，民生码头作为滨水空间的一部分，其改造作为本次空间艺术季的具体案例，展现了如何连接和重整断裂的城市空间，既契合了本次空间艺术季"连接"的展览主题，又展示了黄浦江东岸贯通开放工程和城市更新的成果。

8万吨筒仓乃至整个民生码头的保护性开发，在充分挖掘和展示城市历史文脉的同时，结合码头地区更新，构建一个以艺术展览、艺术品展示交易、文化演艺、创意设计、精品酒店等为主导功能的公共活动区域，形成具有地带性特征和引领区域文化的"民生艺术码头"项目。未来，浦东新区将致力于打造兼具活力与魅力的高品质浦东，建设一批重大文化设施，大力提升滨水空间的活力，而"民生艺术码头"将成为浦江东岸文化集聚带的重要组成部分，一颗闪亮的明珠将屹立在浦江东岸。

Shanghai Urban Space Art Season (SUSAS) 2017 was launched in October. Any visitor of the Main Exhibition may have been stunned by the renovated Minsheng Port: It used to load and unload sugar and grain, and the 80,000-ton Silo, previously the largest silo in Asia, has left few traces of its glorious industrial past. With the recently constructed public spaces on both banks of the Huangpu River, the long-forgotten port has been regenerated, while the silo is chosen to hold the main exhibition of SUSAS 2017.

Shanghai East Bund Investment (group) Co., Ltd initiated a protective renovation project on the silo, trying to preserve its original exterior to the fullest. The exhibition took place on the ground and top floor, integrated to activate the structure artistically. We have attached an escalator to its exterior so that the visitors are directly transported to the top. The light, safe and functional escalator constitutes the silo's organic section, and shapes the waterfront horizon by endowing visitors a grand view of the Huangpu River and Minsheng Port. The renovation of Minsheng Port, part of the waterfront space, is a site project of SUSAS 2017 demonstrating how to connect and reorganize fractured urban spaces. It is consistent with the art season's theme, "thisCONNECTION", and showcases the results of the Connection Project Leading Down the East Bund of the Huangpu River and Urban Regeneration.

Based on revealing the cultural heritage of the 80,000-ton Silo and Minsheng Port, we plan to construct a public activity zone whose primary functions include art exhibitions, showcasing and dealing in works of art, entertainment, innovative design and boutique hotels, and fostering a "Minsheng Port of Art" will help to represent the area and lead the community's cultural development. The efforts are integrated with regenerating the area around Minsheng Port. In the future, authorities of the Pudong New Area will strive to create a more vibrant and charming urban area, constructing a series of key cultural facilities and enhancing waterfront spaces. The "Minsheng Port of Art" will become a key part of the Pudong Cultural Zone, a shining star on the east bank of the Huangpu River.

连接城市空间 以心点燃人们的心
A CONNECTION OF URBAN SPACE, A TORCH TO LIGHTEN

郑时龄　学术委员会主任

ZHENG Shiling　Academic Committee President

城市是人类的化身，城市和城市空间是按照人们心目中的期望所塑造的世界。人们在构造城市和城市空间的过程中塑造了城市的未来，也重塑了自身。城市空间关乎城市的价值取向，关乎城市之人，关乎城市的未来。上海自2015年起举办城市空间艺术季的初衷，是打造一个以展示上海城市空间为平台的品牌活动，宣传和推广城市发展的理念，塑造城市空间，展示城市空间的优秀作品，从而推动城市的有机更新，完善城市空间品质，在城市空间的塑造过程中提倡高品质的设计，创造宜居的生态环境和愉悦的生活场所，让艺术介入空间。让空间品质和艺术感化人们，使人们成为更加高尚、文明、卓越的全球城市之人。城市空间艺术季不仅仅是展览和理论的探讨，也是实验性案例的展示和城市更新的实践，通过城市空间艺术季的建设和展示活动，上海的城市空间越加美好，城市环境越加宜居。

20世纪90年代以来，上海进入了重要的历史时期，城市空间结构和城市产业结构进行了全面的重组。上海在城市转型发展的过程中，面临着前工业时期和工业化时期遗留的问题以及新的后工业社会的问题。城区内仍然存在各个地区之间的差异，在环境、交通、文化教育、公共服务设施和医疗卫生资源方面存在不平衡。上海新一轮的城市总体规划提出了建设卓越的全球城市的目标，建设更具活力的创新之城，更富魅力的幸福人文之城、更可持续发展的韧性生态之城，塑造具有文化地域性、国际都会感的城市总体风貌，历史文化空间与当代功能的空间层积，弘扬地域文化的历史性与全球城市的当代性。

由伦敦市长办公室发起的关于城市文化和未来发展的全球性倡议《世界城市文化报告2012》指出："世界城市在文化方面的重要性，与它们在金融和贸易方面的重要性等同。"上海作为卓越的全球城市，应当是一座国际化的大都市，一座多元的城市，一座拥有无与伦比的艺术和文化活动的城市，一座拥有国际学术和研发中心的城市，一座充满生命力的城市，一座以人民为中心的城市，一座安全和绿色的城市。作为全球城市，也应当是世界级的文化中心，具有持续创造科技成果和知识信息交流功能的文化、教育、科研中心。以此为目标，城市空间艺术季也承担了重要的文化使命。以城市之心去发现心，以城市之心点燃人们的心灵。

以"连接thisCONNECTION：共享未来的公共空间"为主题的2017上海第二届城市空间艺术季，展现了城市空间与建筑和公共艺术。这个主题延续了第一届城市空间艺术季的主题"城市更新"，直面当前城市空间中的问题。我们的城市正面临着多种多样的断裂，空间的、社会的、文化的、历史的、代际等的断裂。长期以来，在城市外部和内部的社会、经济、文化等各种影响力的作用下，地区之间、城郊之间、黄浦江和苏州河两岸之间、人与自然之间都呈现了某种断裂的状态。不断往高空发展的高密度特大城市使人们在空间中的活动缺乏连续性，快速的生活节奏使人与人之间缺乏沟通，各社会阶层之间的隔阂也造成了断裂。为此，本次城市空间艺术季主题的提出是上海城市空间中各种"断点"的映照，将断裂转化为连接和延续，隐喻了城市历史文化的传承和发展。主展览所在的民生码头和8万吨粮仓正处于黄浦江滨江贯通的一个重要的节点，反映了黄浦江45公里滨江贯通的理念和实践，同时也是又一次以主展场的建设引领城市更新的杰出案例。

本次由意大利建筑师博埃里、中国建筑师以及艺术家李翔宁和方振宁联合策展的城市空间艺术季主展览包括了4个主题展和12个特展，全市范围内还有8个实践案例展和6个联合展，以及100多场各类公众活动。城市空间艺术季也越发吸引了社会和公众的广泛参与。

A city is an incarnation of people. As was expected by people, the city and the urban space are established. In construction, people have built the city's future and rebuilt themselves. The urban space of a city is as much about its future and citizens as its values. To set up a platform to present a city's urban space, Shanghai has held Shanghai Urban Space Art Season since 2015 to promote its vision for urban development, establishing an urban space and presenting outstanding works. The organic renewal of the city has been accelerated and quality of the urban space has been improved. High-quality designs have been introduced through the urban space construction to create a livable ecological environment and pleasant living space. Art has been melted into spaces; the quality of space and the sense of art can lead people to become more noble citizens, with principles and achievements. Shanghai Urban Space Art Season will not only provide exhibitions and welcome discussions on theories, but also present experimental samples and offer urban regeneration practices. Through the construction and exhibition of the Urban Space Art Season, Shanghai would welcome a better urban space and more livable urban environment.

Since the 1990s, Shanghai has entered a significant historical period. The space and industrial structure of the city have both experienced full reorganization. Transformative development has to face issues left by the pre-industrialization period and industrialization period that emerged in post-industrial society. Districts of Shanghai still have differences and imbalances in aspects of the environment, transportation, cultural education, public service facilities and health care resources. The new Shanghai Master Plan has set up a goal of establishing a new global city: more dynamic and creative, with a charming humanity and more tenacious with its sustainable development and ecology, to establish an overall characteristic that combines regional cultural features and international metropolitan characters; to accumulate historical and cultural spaces with modern functional space; and to carry forward the historical significance of regional cultures and modernity of global cities.

The World Cities Culture Report 2012, initiated by the London Mayor's Office, has discussed urban culture and their future development, and pointed out: "World cities share the same significance in culture as in other aspects such as finance, trade and so on." As an outstanding global city, Shanghai should be an international metropolis and dynamic city, offering artistic and cultural activities with international academic and research centers. It shouldn't only be a city of vitality, people-oriented, safe and green, but also a city with its own world-class culture, education and research centers, where continuous scientific and technological achievements are made and information and knowledge are shared openly. To achieve this goal, Shanghai Urban Space Art Season has undertaken its cultural mission. The soul of a city is a key to unlocking its mysteries, a torch to light its hidden areas.

Themed by "this CONNECTION: Sharing a Future Public Space", the 2nd Shanghai Urban Space Art Season has been held this year, presenting urban spaces, architectures and public art. Carrying forward "Urban Regeneration", the theme of the first Urban Space Art Season, this time it has faced up to issues in current urban space. Our urban life experiences have developed in various aspects, including space, society, culture, history and intergenerational relations. For a long time, under the influences of inner-city and suburban society, economics and culture, fragmented states have formed among regions, between the city and its outskirts, or between the Huangpu River and both sides of Suzhou Creek. People have been suffered from the fractures with nature. Activities form a lack of continuity for the megacity of a high density that keeps roaring to the sky, and people have little communications with others for its rapid pace of life. Gaps between social classes have also contributed to its fractures. This Urban Space Art Season proposed the theme as a reflection of "break points" among Shanghai's urban space, in hopes of transforming fractures into connections and continuity, by which the city's history and culture would be carried forward. Located at Minsheng Port, the 80,000-ton Silo, the Main Exhibition venue, stands as a joining point of the Huangpu Waterfront Connection, presenting the vision and practice of the 45km Huangpu Waterfront Connection, and displays itself as an outstanding example of leading urban regeneration by virtue of the exhibition venue.

With the work of joint curators, an Italian architect, Stefano Boeri, a Chinese architect, LI Xiangning and a Chinese artist FANG Zhenning, this Urban Space Art Season includes 4 theme-related exhibitions and 12 special exhibitions. In addition to the main exhibition, 8 site projects, 6 joint exhibitions and over 100 theme-related civil activities will also be held in Shanghai. Shanghai Urban Space Art Season has received growing participation from the public.

从公共艺术空间连接未来城市
"thisCONNECTION": FROM PUBLIC ART SPACE TO A FUTURE CITY

斯坦法诺·博埃里 Stefano Boeri 主策展人 Chief Curator

2016年初,我接受了上海城市空间艺术季(SUSAS)的邀请,作为主策展人之一策划2017年的这场盛会。这对我和我的团队而言无疑是件令人兴奋的事情,因为上海是一个讨论未来的理想之地,它代表着未来所有的可能性,包括建筑、城市、艺术、空间、公共交流等众多城市元素。我们设想2017年的SUSAS会对现在、未来以及人们寄予未来的期望产生显著影响。SUSAS能通过创新、交流打造出新一代的公共空间。

公共空间本身是让人们交流思想和情感的场所,如果人们想要并且需要的话,在此过程中,也能感到一丝亲密。我们借SUSAS来思考"未来城市的公共空间是否有足够的资源让人们方便地联系或断联"以及"如何利用想象力去设计亲密环境"这些议题,并在此过程中明白,如何在这个环境里重建人们的实际联系。

然而,这并不容易。每一天,当我们在思考我们的工作时,建筑师、城市规划师、设计师,对于我们每一个从事城市空间规划和改造工作的人来说,似乎不可能不去面对对于物质世界的操纵行为,而非接受我们行为的需要以及现实的存在,这些现实存在带有不同且有时甚至并置的观点。

这种并置或者说是振荡可以用几种方式来定义。

介于分析和与分析并行的项目之间的这种振荡,并不是按照时间顺序和/或因果顺序进行的。我们的角色是要处理、研究和调查包容性领域(设想一个空间未来的无数可能性的好奇心)以及设计的选择性领域,在孤独中,产生一个单一的"封闭"配置(关闭形式、重力、材料、重量的部分)。

在时间概念上的这种振荡,对建筑师而言是一个深不可测的概念:与完成作品的寿命相比,是非常短暂的;与直觉的节奏相比,是非常长久的;与社会风俗的变化相比,是非常迅速的;如果与品味的癖好和舆论的期望相比,则会变得非常缓慢。

我们已经理解了允许这两种建筑体验模式的有用性;在允许两种模式同步进行的同时,就有可能产生意想不到的火花和突触,并提出一种非常特殊的方式来创作我们所称的"项目"。

这些场所既是方法论的挑战,也是我们运营中的挑战,于是就有了我们与李翔宁教授、方振宁先生共同策划的2017上海城市空间艺术季(SUSAS)。今年的展览题目是"连接thisCONNECTION:共享未来的公共空间",我们接到要求,需要考虑截然不同的情况。随后这个主题在三个月内向世界各地的游客们展示艺术、建筑、规划、设计、当代人文等方面的优秀作品。

我们在SUSAS面临的第一个挑战就是为这个盛会想象一个装置构思,以便重新定义上海民生码头的空间。

我们提议了"Radura 林中之境"。此艺术装置首次亮相是在2016年"米兰设计周"期间,当时在米兰大学的Pharmacy of Ca 庭院中展示。

森林内部空地起着关键作用。这些空地被雕刻并组合在一起,它们代表了生物多样性繁殖和增殖的条件。在城市环境中,Radura(一个由木制圆柱组成的装置)具有相同的含义:它是大都市人潮中的一个公共空间原型。它代表着从大都市繁忙的生活中放慢脚步的机会,并且创造了一个混合空间,让男人、女人、孩童和动物们在推动生活的同时,让自己和他人重新联系起来。它的渗透性边界创造了一个同时与外界分离和连接的状态,一个意外的亲密空间,在这个空间里,运动和流量冲淡了它们的力量,木材料创造并放大了与城市噪音相对立的声景。它的形式简单，使其成为一个原始的未来主义空间：Radura 成为一个天线，接收来自未来的信息，并预测其形式，同时将人类与其起源重新联系起来。

就技术而言，Radura 是一个由 350 根圆柱与一个直径为 10 米的木台组成的圆环，其圆柱通过一个高 45 厘米、外径 10.5 米、内径 7 米的圆形木制平台固定在地面上。在其内部，脚踏成为座位。事实上，Radura 是一个公共空间的理念，让人们可以停下来休息、等待。Radura 是开放的，它有一个可渗透的边界，能够被放置在更大的空间中。它是一种城市设备，用以淡化流量和运动的强度。Radura 是与城市节拍相对立的完整空间的体验。

正是因为它的混合性，它代表了一个机会，在自然环境和人类环境之间建立起强大的空间和象征联系。作为森林中的空地，它通过创造一个悬浮空间来促进生物多样性，在这个空间中停止、休息、等待生成和再生的空间。但为了更好地理解 Radura 的空间意义，我们需要退后一步，从地质的角度观察我们的星球。

一方面，Radura 是一个修改空间的动作；另一方面，它为场地和访问者之间的交流和互动开辟了新的可能。因此，我们所面临的第二个挑战是想象一个令人兴奋的景观，同时使人们可以沉浸其中。

与同济大学未来城市实验室一起，我们设想了一些 2117 年上海大都市可能的城市布局情景，因为在不久的将来，海平面的上升将导致使数以万计的居民无法再居住在城市之中。今后，人类很可能在扭转气候变化的当前努力上失败。

由于不断持续的环境危机，不可阻挡的人口膨胀，地球上广泛的城市化和自然资源的破坏，今日的我们事实上已经在为人类生存的世界做打算。我们也需要接受城市行为的质变，以及它们受日常生活的影响而产生的变化方式，我们如何相互联系，如何消费，如何使用物质和媒体，以及设备。

几年前，我提出需要用一种非人类中心主义的理论来处理城市规划和建筑。为了人类的生存，这个理论把我们的行动放在一个更广泛的限制和可能的场景之中。关键在于，只有在更广泛的世界观中，我们才有可能计划地球的未来。非人类中心主义理论不会因此放弃人类的命运，而只是把它置于一种新观念的中心。

转机已经过去，今天我们正面临新的挑战。

在一个研究项目中，我们想象了一个由世界所有城市组成的巨大的聚集地。它将占据 2% 的世界陆地面积，产生 50% 的世界 GDP，以及 90% 的知识与注册专利。与此同时，二氧化碳排入大气的比例将达到 70%。这个城市，33% 是由贫民组成的，35% 的面积被三个国家的城市所占据，它们是印度、中国和尼日利亚。这标志着城市和当代城市形势思维的巨大转变，这会导致建筑和城市设计的不同方法。因此，一个不能回避的关键问题就是环境危机，这个危机伴随着我们时代的艰难挑战：新的城市人口的融合，气候变化对生活条件的影响以及我们的生存。

化石燃料燃烧产生的二氧化碳排放量的增加以及地球温度的升高，都在威胁着人类文明，使生态系统发生前所未有的变化。面对风险的上升或现实的震荡，世界上一些大城市（上海、深圳、香港、迈阿密）沿海地区的民众将遭受海平面上升的影响。

出于这个原因，我们更加不能忽视建筑和城市是能源的主要消费者这一事实，它们也是造成全球变暖的主要原因。关于小城市、都市、城市星球的政策至关重要，这应该是目前我们的责任，这需要让所有人都参与进来。我们作为建筑师，应充当未来的传感器，预测将要发生的事情并帮助制定应对方式和解决方案。我们的责任也是用我们的想法和设计来促进当下，并以某种方式推动和揭示它的潜力和未来，或者说不仅仅是它的未来。

如果我们从目前森林自身大约吸收 35% 的城市二氧化碳排放量着手，就会明白为什么现在城市中的森林是不可或缺的。在与城市造林主题同步的政策中，我们的团队特别采取了三种方法。首先是像我们这样在城市之中进行生物多样性移植的可能性研究。例如，垂直森林的实现。这个位于米兰的第一个原型使我们能够根据对住户、植被和建筑物行为的研究调整我们的观察方式，以便在其他情况下提出同种解决方案的改进版：我们已经在中国和亚洲其他国家，以及在欧洲（我们在乌得勒支、洛桑和地拉那工作）和美国进行拓展。

我们工作的第二个方向是线性和一体化的基础设施建设，结合植被、水文和交通（甚至条例，现在指的是"绿色和蓝色的基础设施"），我们已经把这个方面用在了米兰铁路站场项目、地拉那的总体规划以及我们正在澳门规划的绿化带项目中。

第三种设计方案由森林城市组成。这些都是小型的新建的城市，在水平面和建筑外立面上都有大面积的植被，这是我们主要为中国而做的设想，目的是应对一个会涉及生物多样性、且重点在环境和空气方面的一个新情况，即：人口从农村向城市的巨量流动。。

与实验一样，这同样需要持续的进化，每个实验都需要修改、变形、重新解释，甚至粉饰以证明其实力。

在 SUSAS 中，我们从垂直森林开始，一个树木和人类之间亲近关系的实验：来想象一个由数十个绿色塔楼组成的绿色城市，一个森林之城，可以在红色星球上容纳此前巨型城市流离失所的人口。多年的科幻电影挑起了人们对于火星殖民化设想的多种方法、情况、技术和难点。我们在思考，我们应当怎样去想象火星的城市化进程？我们是否还可以在探索这个主题漫长的投机过程中增加什么？我们借此机会与空间机构的航空航天研究领域专家和工程师进行了讨论，并对增强现实进行了测试。我们想象一颗胶囊能够包含为人类生活创造条件的景观。我们倡议种子银行，它可以把我们的原生态景观带到新的星球。我们提出了一种关闭和着陆的系统（从火星流浪者装置学到的），它可以移动这些巨型建筑物。我们在这个封闭系统中使用带有生物再生的生命支持系统的圆顶：完全隔离，无进无出。一切都将被使用。

最后，我们用 AR 技术通过增强现实的设置揭示了这个想法。通过联系城市居所的未来愿景，等待着去揭示一个伟大的概念，更是一种引人入胜的体验。游客可以使用现场设备或下载应用软件体验 100 年后的在未来星际之中的绿色城市景观。

这是一个刺激的挑衅，但也预示着一个迫近的答案——环境正在为解决自身问题寻求对策。

At the beginning of 2016, I accepted an invitation of the Shanghai Urban Space and Art Season to curate the 2017 event as one of the curators-in-chief. It was exciting for me and my team, as Shanghai is an ideal place to discuss the future, representing all its possibilities, including architecture, urban life, art, space, public exchange, and other urban elements. We envision SUSAS 2017 to be a grand event with a significant impact on the present, the future and on people's hopes. SUSAS helps to envision a new generation of public spaces through innovation and communication.

Public space—that special area where people can exchange ideas and emotions with others, intimating with people's wants or needs. We think about the topic of "whether future urban space has enough resources to connect and disconnect easily" and "How to use your imagination to design an intimate environment" trough SUSAS, all the while trying to understand how to rebuild people's relationships in this unique environment.

However, this is not easy. Every day, when thinking about our work, architects, urban planners, designers, or whoever works in dealing with planning and transforming urban space, it seems impossible to avoid to confront the actions of manipulation of the material world without accepting the need for confronting our actions with the simultaneity of opposing and juxtaposed points of view.

This juxtaposition or oscillation can be defined in several ways. The oscillation between an analysis and a project running parallel, without ever putting them into a chronological and/or causal sequence. Handling the inclusive sphere of research and investigation (the curiosity that envisions uncountable possible futures of a space) and the selective sphere of design that leads, in solitude, to a single "closed" configuration (closed in terms of form, gravity, materials, weight).

An oscillation in the concept of temporality, an unfathomable notion for us architects: short in comparison to the life of a completed work, and long compared to the rhythm of formal intuitions; fast if compared to changing social customs, and slow if placed alongside whims of taste and expectations of the public.

I have understood the usefulness in allowing these two modes of experiencing architecture; while allowing them to act simultaneously, it is possible to create unexpected short circuits and synapses, and this suggests a particular way of creating anticipatory ideas of the future we call a "Project".

Having these premises both as methodological approach as challenges in our operations we approach the theme of the Shanghai Urban Art Space Season (SUSAS) 2017 co-curated with the Professor LI Xiangning, vice president of School of Architecture and Urban Planning at Tongji University, and independent contemporary artist FANG Zhenning. The exhibition for this year is entitled "ThisConnection - Share the Future of Public Spaces" and we were asked to think about diametrically different situations. The theme showed the art, architecture, planning, design, contemporary humanities and other aspects of outstanding works from around the world to visitors in the following three months.

The first challenge we faced in SUSAS was to imagine an idea for an installation for the event, that could redefine the space at the Shanghai Minsheng Wharf.

Our proposal was Radura, an installation born in Milan in 2016, presented in the Court of Pharmacy of Ca' of the "Università Statale di Milano" for the "Salone del Mobile".

Inside forest clearings play a key role. As voids carved into a mass, they represent the condition for the proliferation and multiplication of biodiversity. Inside the urban environment, Radura – an installation made up of a circle of wooden columns- has the same meaning: it is the prototype of a public space for decongestion within metropolitan flows. It represents the opportunity to slow down from a hectic city life, and to push to live while creating a hybrid space where men, women, children and animals can reconnect with themselves and others. Its permeable perimeter creates a condition of simultaneous separation and connection with the world creating an unexpected space of intimacy where movements and flows dilute their intensity and where wood creates and amplifies a soundscape in counterpoint to city noise. The simplicity of its form makes it an essential space that is at the same time primitive and futuristic: Radura is an antenna that receives messages from the future and anticipates its forms, while reconnecting humans with their origins.

Technically, Radura is a circle of 350 cylindrical columns within a 10m diameter, with wood as its columns anchored to the ground through a circular wooden platform, 45 cm high, with an external diameter of 10.5 meters and an inner one of 7 meters – where the footboard becomes a seating area. In reality, Radura is the idea of a public space for decongestion, to stop and rest. Radura is the project of an open space with a permeable perimeter, placed within a larger open space. An urban device to dilute the intensity of flows and movements. Radura is the embodiment of a sound atmosphere counteracting urban beats.

Thanks to its hybrid nature, it represents an opportunity to create a strong spatial and symbolic connection between the natural and anthropic environment. As clearings in forests, it fosters biodiversity by creating a suspended space in which to stop, rest, and wait within a generative and regenerative space. But to better understand the meaning of a space as Radura, we need to take a step back and observe our planet from a geological point of view. On one side, Radura is an action of modifying the space. On the other, it opens up new possibilities of occupation and interaction between the site and its visitors. The second challenge we were involved in was to imagine a provocative landscape in which people could feel immersed. Together with the Future City Lab of Tongji University, we imagined urban placement scenarios for Shanghai in the year 2117, after the threatening possibility of a not too distant future, the rise of water would make the city uninhabitable to its millions of residents. In the future, humanity will have failed to alleviate extreme weather conditions.

Today, we're pushed towards thinking about the world in terms of the survival of humanity, due to the ongoing environmental crisis, the unstoppable population expansion, the widespread urbanization of the planet and the destruction of our natural resources. We also have a need to accept qualitative changes in our urban behaviors, and the way they have been affected by changes in our daily life, how we relate to each other, consume, use our own materials and media, and our attachment to technological devices.

Some years ago, I proposed the need for a non-anthropised ethical approach to urban planning and architecture. Thought for the survival of mankind, this theory placed our actions within a wider scenario of limits and possibilities. Only within a wider world view, the point is that it is possible to plan a better future for our planet. Non-anthropised ethic doesn't abandon mankind to its fate, but simply places it at the centre of a new discourse.

The point of inflexion has been past and today we face a new challenge.

In a research project, we imagined a gigantic agglomeration made up of all the cities of the world brought together. It would occupy 2% of the world's land surface, but would produce 50% of the world's GDP, 90% of registered patents. At the same time, it would be responsible for 70% of carbon dioxide emissions into the atmosphere. In this city, 33% is made up of slums and 35% of its area is taken up by cities of just three countries: India, China and Nigeria. This marks a huge shift in thinking about the city and the contemporary urban situation and leads to a different approach in architecture and urban design. A crucial question that can no longer be avoided is the environmental crisis, together with difficult challenges of our age: the integration of new urban populations, the effect of climate change on living conditions, and our survival.

The rise in carbon dioxide emissions from burning fossil fuels and rise in the earth's temperature are threatening an unprecedented change for human civilization and ecosystems. When facing the rising risk or the shock of reality, millions of people living in coastal areas in some of the world's biggest cities (Shanghai, Shenzhen, Hong Kong, Miami) will be affected by rising sea levels.

For this reason, we can't ignore that buildings and cities are the leading consumer of energy and major contributor to human-caused global warming. Policies on cities, metropolises, and the urban condition of the planet are crucial and today should be the responsibility of everyone. Our role as architects is to act as sensors of the future, anticipating what will happen and helping to form responses and solutions. It is also our duty to tempt the present with our ideas and designs, in a way nudging it into revealing its future potential.

With forests absorbing about 35 percent of carbon dioxide produced in cities, it becomes clear why the current presence of woods in cities is indispensable. Among the policies that go hand in hand with urban forestation, our studio has worked on three specific approaches. The first is the potential in carrying out pockets of biodiversity throughout the city, as we did. One example: the experiment of the Vertical Forest. The one in Milan is a prototype that allowed us to adjust our way of seeing things, based on a study of the behavior of residents, the vegetation and the building, in order to propose an improved version of the same solution in other situations: in China and other countries of Asia, as well as in Europe (we are working in Utrecht, Lausanne and Tirana) and the USA.

The second direction on which we are working is that of linear and integrated infrastructures, combining vegetation, water and transport (even regulations now refer to "green and blue infrastructure"), which we have brought into play in a project for Milan's railway yards, the general plan of Tirana and the green belt we are planning in Macao.

The third scenario of design consists of Forest Cities. These are small, newly-founded cities with large areas of vegetation, in both a horizontal sense and on the façades of buildings, that we have conceived chiefly for China, in order to accommodate a new scenario which involves biodiversity and highlights the quality of the environment and air — the great exodus of the population from the countryside and into the cities.

As with experimentation there is a constant need for evolution. Each experiment needs to be changed, inflected, reinterpreted, and even counterfeited to prove its strength.

In SUSAS, we started from the experiment of The Vertical Forest - an experiment in a new relationship of proximity between trees and people — imagining a Forest City including dozens of green towers. A forest city that could accommodate a displaced population of a former megacity is the red planet. It is provocation fueled by years of science fiction movies that have envisioned a multitude of approaches, situations, techniques and difficulties for the colonization of Mars. We wonder how we could imagine this process of urbanization. What could we add to a long speculative and inventive process exploring this theme? We took the opportunity to discuss with specialists in aerospace research and high-tech engineers from spatial agencies, and we tested augmented reality. We imagine capsules that are able of comprising a landscape to create conditions for human life. We proposed seed banks that could take our native landscape to a new planet. We proposed shuttling and landing systems (taken from the Mars Rovers apparatus) that could move these megastructures. We used Bio-regenerative Life Support Systems domes for a closed system where nothing goes in, nothing goes out. Everything is used.

At last, we exposed this idea using an augmented reality set, waiting to create not only a strong concept but also an engaging experience, a connection through the vision on the future of urban settlements. Visitors can use on-site equipment or download application software to experience the green city view in space after 100 years.

It's a provocative concept, but also a sign of finding out an urgent answer — the environment in search of a solution to its own problems.

超越展览的展览：作为城市的介入方式
EXHIBITION BEYOND EXHIBITION: A WAY TO INTERVENE CITIES

李翔宁　主策展人　LI Xiangning　Chief Curator

纵观历史，尤其是20世纪，一些重要城市空间的发展都与展览息息相关。比如德国国际建筑展（IBA）从1901年的德姆斯塔特开始到1987年的柏林到1993年的汉堡，从"为建筑艺术家提供一个聚落"的口号开始到"内城作为居住场所"到"都市气候变迁"，展览已经从一个单纯展示建筑作品的展览衍生到城市生活、城市更新的介入方式。

在今年的空间艺术季，我们希望将讨论的主题从建筑的范畴拓展到公共艺术的领域，实际上包括环境艺术，以及设计介入空间的方式，包括了建筑、城市、公共艺术、雕塑、装置。而这届的艺术季，缘起于上海今年非常重要的主题——两岸城市空间的贯通——黄浦江两岸有45公里的水岸线的贯通。上海市政府希望将这个重要事件结合进这一届艺术季。所以从这个事件出发提出了一个概念叫作"连接"。在英语单词上我们玩了一个双关，一般读起来像是thisCONNECTION，但是一般会听成disCONNECTION，就是断裂，我们也可以理解为从断裂到连接转化的过程。我们将主题的副标题叫作"共享未来的公共空间"，所以这是我们展览的目的。

thisCONNECTION 的复合展览结构

thisCONNECTION 中的 this 不仅仅强调连接，更是展览结构的复合表征词汇。"t"指代 topology，"h"指代"heterogeneity"，"i"指代"infrastructure"，"s"指代 Shanghai sample。这四个词语在展览结构中形成四大板块。

Topology："公共空间形态"板块主要呈现于一楼筒仓空间与户外广场。一楼的筒仓空间，30个筒仓列为3排，每一个投影面都是12米直径的圆形空间。如何在这样一个历史工业空间中延续公共空间的发展和对于公共空间的讨论，是展览一开始策划的核心问题。而另一方面，公共空间由于新媒体、网络数字、技术和生活方式的改变而呈现出新的面貌，新型和传统形态的公共空间的混合将是未来发展的方向。所以我们将筒仓作为基本单元在展览空间中拓扑变化，通过比较和观察基本空间在物体与媒介两种层面上的差异和转换，最后以建筑、雕塑、装置的呈现来表达和探讨混合多样的公共空间形态。在该板块中，博埃里事务所的《林中之境》创造了在自然和人类环境之间创造强大空间和象征性连接的机会；劳伦斯·维纳的《在混乱中迷失》在筒仓空间中用新材料磁片适应了文保建筑的保护需要，与之前的艺术表达形式迥异，这种不同空间中的适应性回应了topology的拓扑关系。

Heterogenity："社会文化多样"板块通过呈现多样的地域、历史、机制和价值观，把整个展区转变成内含多样社会文化的场所，选择本身所具有的独特的社会文化基因案例来激发上海在全球-地方城市建构过程中的积极思考。北川富朗的《濑户内国际艺术祭》激发了偏远岛屿的文化生态与社会活动，《南京长江大桥记忆计划》用物品的纪念性回顾了历史中的南京长江大桥以及集体的力量，用以探讨记忆作为人与空间的触媒可能。

Infrastructure："基础设施连接"板块除了展示传统道路和桥梁的新的利用方式，还展示互联网技术和新材料带给空间和时间新的维度的连接体验。数字金属、超薄纸板大跨建构和机器木构集中呈现数字化设计与建造所带来的变革，几何形式与结构性能之间得以关联，材料与制作手段之间得以同步，工具、方法与工序之间获得了前所未有的紧密连接。

Shanghai sample："上海都市范本"表达从一种已有的更新范式到另一种范式的转变过程，不管自上而下还是上下结合的更新模式都进入讨论，映射上海这座城市本身所具有的开放精神。城市更新对城市来说是一座连接城市的过去、现在与未来的桥梁。上海城市空间研究与新村研究借助非常在地的研究内容来讨论城市与空间问题。

在对主题的分解而形成的四大板块之外，12项特展从当代建筑、滨水空间整治案例、历史图集等方面来呈现学界对建筑、公共空间以及城市的思考。《拉斯维加斯工作室：来自罗伯特·文丘里和丹尼斯·斯科特·布朗档案馆的影像集》特展创造了建筑摄影、城市研究、理论和本地贸易之间的关系，其研究手法又与上海当代城市研究产生了共鸣，使得一场跨越时空的生动对谈得以连接；《与水共生：世界优秀滨水空间案例特展》探讨了滨水空间成为城市公共生活的发生地的可能性，通过诸多世界滨水空间优秀案例的展示，关注了后工业城市滨水地带的再生与激活和气候变化下的弹性水岸的设计。这些思考都给上海这座城市的滨水公共空间的再生带来新的思考与参照。

thisCONNECTION 所讨论的未来公共空间的形态

城市公共空间并非仅是单纯物理学意义上的地理空间，还是公众广泛参与的公共领域。如何让公众在城市公共空间生产中扮演重要角色，是未来公共空间治理和良性发展的重要动力。民生码头筒仓及黄浦江滨水岸线作为本次空间艺术季的承载空间，建筑改造与景观改造以具体改造案例的身份出现，其公共空间形态的呈现是展览的有益尝试。建筑师柳亦春在建筑改造的过程中，斡旋于工业建筑本身的特质和建筑保护的框架之内，思考对工业建筑遗产的有效性利用并使之适应文化建筑的需求。建筑改造与展览布置同步进行，在展览场地我们可以同时看到建筑工人、布展人员、艺术家、策展人共同工作，这种共同参与的模式在最后的布展过程中渲染了尤其紧张的气氛。展览本身也是公共空间的一种体现形式，我们讨论公共空间，也讨论参与文化事件的角色与向度。而对于建筑师来说，这一场声势浩大的改造本身已是一件最有力量的作品。

民生码头与筒仓空间的改造以构筑开放的滨水平台为目标，以真正提升黄浦江两岸开放空间的潜在价值，并以此促成更多有着相似"连接"性的未来公共空间。正如主策展人博埃里所号召的：我们期待国际国内艺术大师、建筑师、规划师、摄影师、舞者、音乐家、作家、诗人、科学家、社会学者以及人类学家等齐聚一堂，通过跨领域和学科的艺术呈现，以及公众参与的系列文化活动，让艺术和专业工作者与市民一同在民生码头这个曾见证了上海历史的地标重塑公共空间的未来。

History of the 20th century indicates a close relationship between the development of some important urban spaces and exhibitions. Internationale Bauausstellung (IBA), which started in Darmstadt, 1901 with "Artist Colony" as its theme, has extended from a show of architectural works to various interventions into urban life and city renewal, as demonstrated by its later slogans like "The Inner City as a Place to Live" (IBA'87, Berlin) and "Urban Climate Change" (IBA '93, Hamburg).

In the Shanghai Urban Space Art Season (SUSAS) this year, we hope to lead discussions beyond architecture, to involve public arts, environmental art and numerous ways that design may enter spaces covering buildings, city planning, public art, sculptures and installments. SUSAS 2017 is inspired by a critical Shanghai project in the same year: the Connection Project along the 45-kilometer waterfront space of the Huangpu River. The municipal government of Shanghai intends to incorporate this event into the art season, so we have proposed the concept of "thisCONNECTION". Here is a pun for hearers are inclined to perceive its pronunciation as "disCONNECTION", or fracturing. The theme may be interpreted as a process from disconnection to connection. Supplied by the sub-theme "sharing a future public space", we have expressed our aim designated for this exhibition.

The Complex Structure of "thisCONNECTION" exhibition

"this" in "thisCONNECTION" is an acronym signifying Topology, Heterogeneity, Infrastructure and Shanghai Sample, all of which are the four sectors of exhibition.

Topology: Most items of Topology are placed on the first floor of the Silo Space and the outdoor plaza. Thirty silos are arranged in three rows, each projecting a circular space with a radius of 12 meters. From the very onset, a core curatorial issue is how to present a discussion of public spaces and their development in a site with an industrial past. Furthermore, public spaces have taken faces never seen before thanks to new media, the Internet, technology and changing lifestyle. The future of public spaces lies in mingling both the old and new. That's why we chose silos as the "building blocks" of the exhibition space. They are subject to topological alterations. In comparing and observing differences and conversions, as objects or mediums, of basic spatial forms, structures, sculptures and installments are used to express and explore a form of public space featuring diversity. "Radura" by Stefano Boeri, offers a chance to establish activated spaces and symbolic connections between natural and artificial environments, and "Lost in a Shuffle" by Lawrence Weiner, in adaption to the conservation of historical sites, uses new magnetic materials to present a strikingly divergent form of art, are examples of this sector. In particular, Weiner's work echoes the mathematical character of topological relations with its adaptability in various spaces.

Heterogeneity: By presenting diversified regions, histories, mechanisms and values, this sector makes itself a place of social and cultural pluralism. Cases inheriting unique cultural and social genes are carefully selected to inspire positive thinking about Shanghai's role in urban development caught between globalism and locality. Cultural ecology and social activity on a remote island are evoked by "Setouchi Triennale" by Fram Kitagawa, while the "Memory Project of the Nanjing Yangtze River Bridge" shows memorial objects regarding the history of a bridge and its consolidation, so as to explore the possibility of memory as a mediating catalyst between people and space.

Infrastructure: Besides demonstrating new ways to utilize conventional roads and bridges, this sector is dedicated to newly-connected experiences, spatial or temporal, enabled by the Internet and new materials. "Digital Metal" , "Shells with Thin Sheet Materials" and "Robotic Timber Construction" highlight transformations made possible by digital design and construction, which correlates geometry and capacity, material and manufacturing, binding tools, methods and procedures closer than ever.

Shanghai Sample: This sector is demonstrating a paradigm change of urban regeneration. Top-down and bottom-up strategies are both considered as reflections on the open mindset of Shanghai. Regeneration, for a city, is a bridge that connects its past, present and future. Among others, "Shanghai Urban-Space Research" and "Study of New Villages" are particularly localized works on cities and spaces.

Twelve special pavilions, which are not part of any of the four main sectors, demonstrate scholarly thoughts about architecture, public spaces and cities, covering contemporary buildings, environmental improvement of waterfront spaces, historical pictures and more. "Las Vegas Studio – Images from the Archives of Robert Venturi and Denise Scott Brown" initiates a relationship between architectural photography and urban research, academic research and local trade, in a manner that echoes urban studies of contemporary Shanghai, enabling a conversation that traverses time and space; while "Living with Water: World Extraordinary Waterfront Space Cases" discusses multiple possibilities of making waterfront space a place for the public to walk about freely, demonstrating examples of waterfronts from around the world, focuses on revitalizing post-industrial waterfronts and the design of resilient waterfronts in response to changes in climate. All these bring new insights and references into renewing Shanghai's public waterfronts.

Forms of Future Public Spaces

Urban public spaces are not physical locations, but public domains for people to openly interact and engage. A primary dynamic for the proper governance and development of future public spaces is to motivate the public to play a role in creating such spaces. Silos of Minsheng Port and Huangpu River Waterfront, as chief venues of the SUSAS 2017 exhibition, are also practical cases of architectural/landscape renovation, making due contributions to demonstrating various types of public spaces. Yichun Liu is the architect of our project. As he assists in the architectural renovation project, his position revolves around compromises and counterbalances between two distinct aspects of the same building: as an industrial facility and a conserved structure. He is thinking about the usefulness of industrial heritage and how to take advantage of this as per the demand of a cultural institution. Renovation and exhibition occurred at the same time. Construction workers, exhibition staff, artists and curators were working together, and this model of involvement reached its climax towards the end of the set-up. Exhibition by itself is an expression of public space. When we talk about a public space, the roles and dimensions of cultural events never leave our mind. Yet for the architect, this renovation project is already a work of divine force.

The renovation projects of Minsheng Port and Silo Space aim to construct an open waterfront platform that enhances the potential value of public spaces along the Huangpu River, and facilitates more public spaces with similar "connectivity" in the future. As what Stefano Boeri, the curator-in-chief calls for, we hope international artists, architects, planners, photographers, dancers, musicians, writers, poets, scientists sociologists and archeologists work together, presenting the art through interdisciplinary methods, with series of cultural programs participated by the public, to reconstruct the future of the public space of Minsheng Port, the landmark historical architecture, with artists, professionals and citizens' witness.

场所决定形态：SUSAS 2017 艺术介入计划

PLACE DICTATES FORM: ART INTERVENTION IN SUSAS 2017

方振宁　主策展人

FANG Zhenning　Chief Curator

两年一次的上海城市空间艺术季（简称SUSAS）是上海迈向"杰出城市"的重要布局，本次上海城市空间艺术季的主题为"连接this CONNECTION：共享未来的公共空间"，而2017年第二届SUSAS现场实况，为这一布局的实效做了一次检验。

定调

上海同中国其他城市一样，是近30年以大规模建设为特征的快速扩张型城市，在经过极速发展之后，就面临城市转型的问题。而SUSAS就是在这一重大历史转折时，应运而生的大型城市活动。两年前，我们看到它的定调是为了起到先导作用，需要进一步激活城市空间。于是怎么激活？从哪里入手？这些就成为更为具体的话题。SUSAS的先锋性和实验性体现在中国其他城市里没有同类活动，上海这座中国的先锋城市自己也未曾举办过。从结构上看它是两年一次，可它又区别于通常的双年展，因为双年展这种形式无论在上海还是在其他城市都有，但主题大都是艺术和建筑，而SUSAS则把概念提升到城市空间艺术的层面。

这个活动的句型实际上是由四个单词复合而成的，即"城市""空间"和"艺术"，最后的"季"是对前面三个意思的统合。"城市"是一个由众多人交流和交际而逐渐形成的群居形态；而"空间"则是人和人、人和物、人和自然之间的平方距离，有了这个空间人们才可以活动，所以"共享"这个空间是非常重要的，因为要达到共享的状态，就需要有伦理、规则和自觉的约束；而"艺术"是提升人的素养、挖掘潜能、启发人的创造力的一种方式，艺术介入空间和参与到城市规模的活动，则可以使城市空间和生活质量升级。所以，SUSAS是中国城市化进程中史无前例的实验，通过三个月的活动获得意想不到的反响。上海的实验必将影响全国其他城市和地区，笔者想从主策展人之一的角度，来谈谈艺术在这次SUSAS活动中的作用和角色。

艺术激活城市死角

上一届SUSAS的主题是关于"城市更新"的思考，而新一届SUSAS加入"艺术"的内容，成为上一届的改良版。是对于"城市为什么需要更新？城市为什么需要艺术？"的思考。

城市是一个肌体，是具有吐故纳新、新陈代谢能力的有机系统。"吐故"就是由于自然的分化造成功能的老朽，因而必须替代和清除的部分；"纳新"就是吸收和接受新事物、新科技出现的现实，提高城市这一肌体的效率和运转功能。由于城市像一个肌体，有着吸收和排泄的循环功能，所以不能只关注吸收和纳入，还要关注它的废弃物，比如废弃物和垃圾处理等，保持城市的气脉和精气才能保持循环。这些属于肌体的部分，既然是肌体，就有头脑的部分。头脑是意识形态和创意思维，艺术就是城市中最为活跃的分子。所以，世界各地多有这样的案例，那就是城市中沉睡的角落常常被艺术所激活。

位于上海浦东新区民生码头的8万吨筒仓，已经结束了它的工业使命，在沉睡了十年之后，第二届SUSAS把主展览定在这里。当高达48米的筒仓重新亮起灯光时，笔者感受到了时间的沉淀。

"筒仓"的全称是"散装粮筒式储存仓库"。一座天桥连接着上海浦东新区民生码头上的两座筒仓，一座是在20世纪70年代建造的4万吨级筒仓，另一座建于20世纪90年代，储存量8万吨，建造物的规模和仓储量均为当时远东地区之最，所以上海不可多得的工业遗产建筑。建筑之所以设计成圆筒式，是因为圆筒式不会留有死角，相对于其他形式可承载力最大。从20世纪70年代投入生产，到2007年正式停产退出历史舞台，这两座筒仓承载了上海乃至全国的粮食供给，可谓功不可没。我们要做的就是在这组建筑被正式改造再利用之前，组织一个意在激活筒仓的艺术季。

场所决定作品

在中国不缺少公共艺术，缺少的是有品位的公共艺术，无品位的作品即使投放到公共空间里去，也不会起到提高观者审美和让人共享的作用。所以，把艺术纳入公共空间，使其成为"共享未来的公共空间"，是要具体到每一件作品，考量每一件作品与环境是否有着契合的关系。2016年8月，笔者开始参与 SUSAS 的投标活动，它和笔者接受上海陆家嘴滨江金融城公共艺术项目的评委之责是前后脚，这两个项目从内容到地理位置上都非常接近，正好用上了笔者过去在日本参与制作公共艺术以及对这方面的研究和实践的经验。

笔者是在20世纪90年代初开始接触到城市公共艺术，那时候只是从艺术的角度考察和报道日本在这方面的活动。日本的策展人、媒体和公共活动都开始围绕都市更新和再开发的内容，杂志也开始出版艺术、建筑和空间的特集。由于笔者也从事极少主义艺术作品的创作，所以空间性的作品是不是可以向城市空间伸展？这成为笔者新的思考方向。笔者既非学雕塑出身，专业也不是建筑学，所以无论是构思，还是使用材料以及手法，都与雕塑和建筑没有直接关联。在这里为什么提出建筑的问题？是因为笔者在1992年左右开始对建筑抱有兴趣，所以创作的公共艺术作品大都是使用建筑材料和构架。

从传统的雕塑转换到公共艺术不是一件容易的事情，二者的不同之处在于是先做好作品放到空间里？还是根据空间去制作作品？公共艺术的规则是场所决定作品。这次笔者负责策划的筒仓里的七件作品，257库的五个特展以及室内外的四件（组）公共艺术都从这个原则出发展开。

艺术介入的案例

T6 / 劳伦斯·维纳 / 混乱中迷失

劳伦斯·维纳是世界著名的美国观念艺术家，能够请他出山参加这次的艺术季，不是一件容易的事情。能够邀请到年事已高的维纳，主要是靠笔者和他近二十年的交情，看来策展人和艺术家之间的个人关系决定了参展作品的段位。

"LOST IN A SHUFFLE"是英文中的一句俚语，无论是字面还是隐喻含义，它均可以用于形容人在某个情境下不堪重负，或者是人故意将自己置于会迷失的情形下。圆周运动既是一个平面上的动作，也是一个文字内容的图像化表达，用以赞美迷失的表象下循环往复的本质。在这种方式下，维纳提出了文字与图像的关系。在维纳的叙述中，所有的表达都经过仔细推敲，并且以几乎跳脱出情境的方式展示出来，带动观者自行解读作品的意义。

在筒仓内安装维纳这件作品的难度超过想象，回顾一下如何在漏斗型的筒仓锥体部分固定作品的文字是让人十分感慨的。首先从维纳的作品可以看出艺术家本人是利用空间的高手，他不像其他艺术家那样习惯性地利用地面、墙壁、柱子以及由八根柱子围和的空间，而只使用了距离地面5m到10m之间的漏斗型仓壁外壁，于是如何固定这些巨大的文字就成为问题。由于筒仓属于被保护的文物建筑，所以它的表皮任何部分都不允许改变，既不允许使用黏液固定物体，也不可以除锈，那么如何把文字固定在金属的仓斗上？主策展团队想出了用软磁片来解决固定问题，即先在软磁片上印刷字体后切割，结果到现场之后发现还是粘不上，于是我们又在字母的边缘用黑磁铁钉，最终解决了下坠问题。艺术家在安装作品之前，发来了详细的色标和安装尺寸图。当笔者把修改后的安装作品方案告知维纳并等待消息时，意外地获得他的赞扬，他对这样的功夫非常满意。这次实践也为维纳今后在类似的环境条件下做作品提供了新的固定方式。而这件作品确实受到广泛的赞誉，并在各种媒体上有着很高的曝光率。

H5 / 吕越 / 蝴蝶夫人

吕越是服装设计师，服装实际上就是最贴近人体的"建筑"，她的作品的灵感来自中国民间印有蝴蝶的蓝印花布，当这些蝴蝶从用蓝印花布缝制的旗袍和服饰中"飞"出来时，真的是美轮美奂，有着梦幻般的感觉。印有蝴蝶的蓝印花布与布料刻成的蝴蝶进行互动，蓬裙的弧形与鸟笼相吻合，把厨娘的围裙和大摆礼服裙进行不协调拼接，加之不同年代的旗袍，创造出淑女与厨娘、自由与禁锢、中国土布与西式裙撑、手工染与激光雕刻、平面与立体、过去与现在、阴与阳、虚与实的对比，而这些看似不相干的东西似乎又显现了相互支撑的和谐。和漏斗状筒仓对应的是地面上一个直径为5m的白色圆台子，四边刻有年代，这些年代和围绕的圆台外高挂的旗袍相对应，和筒仓与圆台形成360°的圆周运动。最初看到这个方案时，笔者想起仰韶文化出土的彩陶盆里作圆周起舞的原始群舞。现在看来，这件充满魅力的装置，确实与圆形的筒仓构成一种静止的圆周形态。

H5 / 王昀 / 空间的边界

王昀是建筑师，但是他已经不甘心于建筑师的身份，所以他的展示方式是从聚落调研、建筑设计、原型研究、抽象提炼演变成绘画，属于一种文本展示。在同样高度的桌面上，分了七个部分，内容为：音乐与建筑、书法与建筑、聚落与建筑、园林与建筑、斗拱与建筑、废物与建筑、自然与建筑，以及21个空间建筑的实际项目，而抽象绘画作品则是他在建筑领域长途跋涉之后的实验。王昀对大家在同一类型的筒仓下展示作品易雷同有所警觉，所以就采取万变归静的手法，即水平无变化，这种极简的展示方式和垂直的柱群形成筒仓简洁明了的构成，达到了场所和形态之间的平衡关系。

H5 / 张永和 + 非常建筑 / 寻找马列维奇

现在我们看到的张永和 + 非常建筑的方案，和最初的方案出入很大，笔者认为初始的方案没有达到和空间默契的程度，于是提出一个要求：可不可以设计得灵动一些？最终方案达到了预想的效果，这个方案给观众提供了可以互动的空间，由于六个可旋转底座和各个不同的取景窗口组成的互动装置，每个都可以进行360°的旋转，所以当观众寻找到相应的几何图形目标之后，就会让取形器的姿态变换各异。

观众可自由利用取景器在空间中寻找与马列维奇有关的几何图形目标，每个取形器分别对应一个几何图形，其中五个对应环境，分布在柱子和筒仓上；一个对应的互动者本身，在一顶可供观展人佩戴的帽顶上。取形器与图形之间并不是简单的对应：取形器的位置和视角会使被观察的图像产生形变，因此设计时对几何图形进行了校正，比如通过方形取景框仰视，会发现和柱上的一个梯形对应了；又由于远距离透视，通过长条形取景框会发现与延展到三根柱子上的长条形对应了。在其他取形器上均可以得到类似体验，观展者在搜寻的过程中有机会对场地有更深入的体验。

那么为什么要寻找马列维奇？筒仓是工业建筑的经典，而马列维奇的至上主义是俄罗斯工业时代的产物，这组装置作品是让参展者在寻找马列维奇几何图形的过程中，体会到艺术和工业美学之间的关系。

T5 / 方振宁 / 凝聚

这个以多种立体几何形组成一组的大型空间装置，是利用257库在特展区以外剩余的空间，借助一些不可移动的与建筑室内相关的设施，也就是对观赏有障碍作用的东西，将其粉刷改装成艺术装置，统称为公共空间中的艺术装置"凝聚"。其中最大的一件为边长约5m的立方体，是在一个原本3m

高的小仓库的基础上加高完成的，其中除了两条短边底边为 4.8m，其余的边长均为 5m。这件装置细微的比例来自列维奇 1915 年创作的红色正方形，这些装置配备在马列维奇视觉年表特展周围，意在暗示其与至上主义在文脉上的联系。场内还有 5m 见方的黑色正方形竖立在一侧——这原本是为了挡住一扇 5m 见方的巨大的仓库用推拉门，但笔者不愿放过任何一个可以把它变为艺术品的契机。

我们把表达空间的语言凝练为几何形，让人想起立体主义这一体系的最终形态，"凝聚"这个概念是向至上主义表达敬意，是空间至上主义的当代版，装置作品《凝聚》实际上是由橘红色不规则立方体、横长三角柱、黑色十字形和黑色正方形等四件装置共同构成的。

P10/ 方振宁 / 马列维奇视觉年表
作为 20 世纪最重要的艺术家之一和至上主义的创始人，卡济米尔·马列维奇（1879—1935）对 20 世纪的全球艺术产生了不可估量的影响，这种影响是在我们对现代艺术流派的深入研究之后逐渐显示出来的。与马列维奇同时代的先锋艺术家不乏其人，但为什么只有马列维奇一人有那么大的影响力？这是因为他的作品的革新性和精神性。他不只在俄罗斯，而且在是世界艺术中形成了自己独特的符号体系，这些符号被他的弟子们发扬光大，并在历经半个世纪之后影响了美国极少主义艺术流派的诞生。编辑这套视觉年表，是在对马列维奇的资料收集和研究长达 30 年之后的成果，也是大数据方法论的结果。为了纪念马列维奇的至上主义宣言发表一百周年，2014 年笔者在北京策划了"马列维奇文献展 / 1879—1935"，这是中文圈首次系统总结和梳理马列维奇的至上主义遗产。现在的视觉年表是在那次展览基础上的增订版。可以说 257 库的整体空间布展都是空间至上主义的实验。

P11/ 方振宁 / 万象
《万象》好像是纸上美术馆，这是一个和观众互动的消费型策划，没想到获得了意外的成功。这个以"万象"为主题的策划，是向十几名艺术家、建筑师和策展人提供一个和观众互动的机会，最后我们定了 18 种招贴，在 30m 的空间里一字摆开。这个互动方式非常简单，就是给每位参加者一个尺寸定为 635mm×965mm，即一张全开纸大小的展出发表机会，把他们的作品印刷在这张全开纸张上，每种印刷总数为一万张，把 18 万张招贴整齐地排列在会场内部，这样闻讯而来的观众需要走到展场的尽头才能取得到招贴，参观者可根据自己的爱好随意带走。我们把它看作是一种特别的传播和消费方式，也可以看作是一种行为艺术。开幕后仅两周，大部分招贴已经被拿走一半，在一个半月的时间里，18 万张招贴已经被取光。

展览即装置

SUSAS 主展览分两个会场，一个是 8 万吨的筒仓，另一个是面积超过 3 000㎡ 的 257 库。从建筑的格局和体量上看，257 库显然没有办法和高 48m、一层面积超过 5 000㎡ 的筒仓相比；257 库的特点是，平面空间比较单一，可使用高度达到 11m。在这样一个空旷的空间里，如何展示就成了棘手的问题。笔者觉得，策展和布展都需要战略性的思考，无论是概念的提出还是对空间的驾驭，特别是面对一个工业建筑，无论是外观还是室内空间，都需要最大限度地保留工业建筑原有的风貌。所以在展示的过程中，尽可能不要对原建筑的硬件设施进行遮挡，我们会根据建筑原有的梁柱框架结构来构思展示的方式，在这里充分体现了"场所决定形态"这一布展概念。

以下是这次布展的五项要素：

1）展览设计是一项功能性和目的性很强的创作，它需要内容和形式的高度结合。但是，笔者不想掉进展览这个行业通常的展示方案的陷阱里，笔者想尝试的是，展览本身和它的视觉总体上应该是一件艺术装置。

2）我们的展览设计不设立笨重的、遮挡视线的展墙，而是改用轻质的布材料来解决展示载体的问题，在设计悬挂这一环节时，笔者采用的是最日常晒衣挂布的方式。这样既提高了布展和撤展的效率，也符合笔者一直坚持的生态布展的原则，让布展的垃圾减少到零。

3）我们让展示版面悬在空中，这样的展示方式，可以让视线自由穿梭，整体灵动。

4）利用室内原有的工业设施，稍加改造将它变为展览可利用的硬件部分，比如将原小仓库改成立体装置，利用原有铁杆梯子作为展览画册的平台等。

5）我们增加了悬挂式展览挂件，这一设计可说是一种冒险，因为我们的经验不足，而承受展品重量的方管所需的尺寸需要一定的经验才能准确预估。我们设计了以边长为 9m 的正方形为原型的四个九宫格形式的悬挂式展架，分别展示四个特展。而悬挂吊杆的高度为 8m，为了减少视觉上的体量，当然是越细越好，最后的技术数据是，垂直方管的边长为 2.8cm，上方横向方管的边长为 3.2cm，用这样的设计来解决悬挂和支撑问题。当然，一根这么细长的方管还需要在它的上部增加一根横向的加强管，整体采用了横搭的方式，而不是挂钩悬挂，从而完美地体现了立方体的九宫格美学。从悬挂顶部到地面高度为 11m，而展览内容的版面控制在 1.2m 的幅面内，可见我们实际才使用了十分之一的面积。用巨大的空来衬托极少的功能空间，这就是"无用之用"。

虽然笔者有着十年的策展经历，但是参与 SUSAS 2017 这样的大型城市公共空间的策划活动还是首次。我们通过艺术进入空间和激活城市死角的方式，可能是城市更新和再生的有效手段，它已经带来明显的社会反响。无疑，它会成为同行和其他城市中同类型活动不可忽视的参考。但是，任何一个成功的案例，都不可能被简单地复制，每次策展都会面临新的问题，都是一次新的冒险。而所谓的"场所决定形态"就是场所决定艺术作品生成的形态，对艺术介入城市公共空间来说，这一点是不变的。

Shanghai Urban Space Art Season (SUSAS) is a major addition of Shanghai's grand plan to becoming an "outstanding city", and two years after its debut, SUSAS 2017 is a sound testament to the themed plan "thisCONNECTION / Sharing a Future Public Space".

Pitch Settled

Rapid growth and large-scaled constructions have been characteristic of Shanghai in the last three decades, like many other Chinese cities, yet the trend is bringing out the issue of transformation nowadays. SUSAS is a major urban event that comes with this historical change. Two years ago, its pitch was settled as a pioneering one with the mission to further activate urban spaces. Specific topics are now under way: how to activate, and where to start? SUSAS is an avant-garde, experimental project, a first for Chinese cities, including the fashionable forerunner Shanghai. What sets SUSAS apart from other biennales, which are not uncommon in Shanghai and other cities, is its focus on the arts and urban spaces, instead of decontextualized art pieces or buildings.

Apart from location, SUSAS comprises "Urban", "Space", "Art" and "Season", the last of which overarches the other three elements. "Urban" life is about numerous individuals converging through communication and socialization, while "Space" exemplifies the squared distances between people and objects, nature or other people. Space sharing is significant, and morality, rules and spontaneous self-constraints are necessary to this end. This is where "art" comes in: a way to both educate people and inspire their potential or creativity. Furthermore, the quality of urban spaces and common daily life is enhanced as art is embedded into cities or citywide events. SUSAS is an unprecedented social experiment in China, growing over three months, and guaranteed to influence other Chinese cities or regions. As one of its primary curators, I would like to discuss the roles of art in SUSAS.

Activating a City's Silent Corners

Based on "city regeneration" of the last SUSAS, we have added "art" into the theme this year. Why regenerate? Why art?

A city is a metabolic organism that eliminates its decayed or redundant parts as it naturally endures and diversifies, and accepts what modern technologies and new realities have to offer for the sake of efficiency and functionality. As the organism metaphor goes, a city doesn't finish its cycle by absorption alone, but generated waste has to be taken care of, for example, litter and garbage, so that nutrition and oxygen flow freely. Each body has a head, and the head of a city is ideology, creativity and art, its most energized element. This frequently happens in silent corners around the world, a restrengthening of their power and voice thanks to art.

The 80,000-ton silo in Minsheng Port, Pudong New District had remained drowsy for a decade after the end of its industrial heyday, before it was chosen to hold the main exhibition of the second SUSAS. When the structures were lighted, I felt showered by a condensed past.

The two silos of Minsheng Port, or "cylindrical bulk-grain warehouses", are connected by a bridge overpass. The 40,000-ton silo was built in the 1970s, and the 80,000-ton silo in the 1990s. In terms of size and capacity, they are one of the largest silos in Far East, making them a unique set of Shanghai's industrial heritage. Their barrel-like, weakness-free shape was chosen for carrying as much strain as possible. From the 1970s to their retirement in 2007, they were used for Shanghai's food supply and other parts of China. Our mission is to launch an activating art season here before they are repurposed.

Place Dictates Art

There are too many public art pieces, yet too few that stand out. Art in bad taste has no place in education or public sharing even when placed in public. For anyone with "sharing a future public space" in mind, care should be taken into consideration about the quality and contextuality of an art piece before situating it into any public space. In August 2016, I began to enter the bid of SUSAS. At around the same time, I was accepted into the evaluation board of Public Art for Lujiazui Harbour City. The two adjacent projects are thematically related; moreover, my career in making public art in Japan, and other academic or practical experiences, proved a helpful cause.

I began to learn about urban space art in the early 1990s. At that time, a wave of city regeneration and redevelopment was emerging among Japanese curators, media and public events, and special issues regarding public art, architecture and spaces were released. I was observing and reporting this from a relatively constrained, art-centered angle. I was a minimalist artist, so I began to ponder the possibility that spatial art should extend to urban spaces. Although I had no background in architecture or sculpture—nothing of my work was directly related to architecture, be it concept design, material or implementation—I proposed an architectural statement because I was interested in architecture circa 1992. That's why most of my public art has been using materials and structures typical of buildings.

It is no easy task to transform a traditional sculpture into public art. Should art be built and then transplanted into a place, or should place dictate how art is composed? The latter is the rule in public art. The seven pieces in silos, five special exhibitions in the 257 Factory and four outside pieces (sets) that I curate all follow this rule.

Cases of Art Intervention

T6/ Lawrence Weiner/ Lost in a Shuffle

Lawrence Weiner is an American artist who dwells on ideas. Given his age and international reputation, it is fortunate that he accepted the invitation to SUSAS 2017, not the least because of our personal acquaintance over the last two decades. The quality of an exhibition is determined by the friendship between its curator and the artists.

"Lost in a shuffle" is an idiom that refers to when one is either overwhelmed in a situation or if some individuals expose themselves to an experience that may carry them away, literally or metaphorically. The circular notion is a graphic gesture and image of textual content, complementing the circular movement of shuffling. In this manner, Weiner suggests a relationship between text and imagery. All phrases in Weiner's statements are carefully chosen and presented seemingly out of context, with meaning perceived by each individual viewer.

It suffices to recall the unimaginable difficulty to install Weiner's work in the silo by reviewing how we managed to print the text onto the silo's "funnel". The work testifies to the artist's masterful manipulation with spatial elements, using only the funnel's outer surface, five to ten meters above ground, for the text instead of supporting them with the ground floor, walls, columns or the limited space, as other artists would customarily do. The text is huge, which makes for a tricky issue. The silo is also a protected heritage, which means neither gluing nor rust cleaning is allowed on the entire surface. This only adds to the challenge. Eventually, the curators came up with the idea of soft magnetic panels. As a start, we printed the text on the panels, cut the panels accordingly and tried to attach them to the metal funnel, only to be met with failure. Our next plan, which proved successful, was to bury black magnetic nails around the panel's edges so that they wouldn't fall. Before installing, the artist had sent us detailed specifications and palette, so we were astonished when we were praised by Weiner when he was informed of our revised implementation plan. What we did may have added some hints to the fixation of Weiner's other future works. Concerning the work, it has become growingly popular and been applauded by the press.

H5 / Lyu Yue / Madame Butterfly

Lyu Yue is a fashion artist, and for her, costumes are the closest "architecture" to the human body. Her work is inspired by traditional Chinese printed blue calico fabric with butterfly patterns. It feels dreamy, as the beautiful creatures "fly" from the printed qipao fabric.

In this work, the fashion artist chooses printed blue calico to interact with the fabric-made butterflies and creates the silhouette of bouffant gown like a birdcage. Meanwhile, she creatively matches a female chef's apron with a ballgown. The work is also accompanied by several different eras of the qipao. Some opposites are in the work: masters and servants; lady and female chef; freedom and imprisonment; Chinese hand-woven cloth with Western crinoline; hand dyeing and laser engraving; plane and stereoscopic; past and present; yin and yang; fiction and reality. These opposites show a harmony between them.

The qipaos are hung from a white circular pavilion, whose diameter is five meters, right below the funnel, with areas on the edges signifying when the qipao next to it was made, forming a 360-degree dynamic with the silo and platform. The first impression of this plan reminded me of a pre-historic group dancing in circles shown on painted potteries of the Yangshao Civilization. In retrospect, this charming installment constitutes a static circular composition with the round silo as part of the bigger picture.

H5 / WANG Yun / The Frontier of Space

Wang is an architect with ambitions beyond building techniques. His paintings are abstracted and evolved from settlement surveys, architectural designs, prototype research, all somehow textual. The exhibition is set on an isoheight platform and includes seven parts: Music and Architecture, Calligraphy and Architecture, Settlement and Architecture, Garden and Architecture, Tou-Kung and Architecture, Architecture and Nature, and Architecture, Junk, as well as actual projects of 21 space buildings. His abstract paintings are experiments of a seasoned architect, aware of the fact that viewers are vulnerable to a monotone impression with a series of works of the same genre in the same silo. He took a radically uniformist and minimalist approach that created a clean picture with vertical columns that struck a perfect balance between place and form.

H5 / Yung Ho Chang & FCJZ / Malevich Finders

Yung Ho Chang & FCJZ's plan we see now is rather different from its first draft, which I believed wasn't concordant with the space, and suggested the designers be more liberal. The final plan offered exactly what I was expecting: six devices, composed of 360-degree rotatable bases and distinct apertures to capture views, and named "form-finders". Viewers might witness a myriad of form-finders and postures when they try to find geometric figures as dictated.

Viewers are free to find geometric figures related to Malevich in the space with form-finders, of which five are placed on columns or silos, and correspond to the environment, while the last one represents an interacting subject as it finds itself on the top of a hat. The correspondence between the form-finders and the geometric figures is not straightforward. The geometric figures would be distorted due to the position and viewing-angle of the form-finders due to perspective. For this reason, the designers have adjusted the geometric figures. For instance, by looking up through a square shaped form-finder would only find it coinciding with a trapezoid-shaped target. Another example would be due to a long-distance perspective — viewing through a bar-shaped form-finder — to sense the coincidence of its aperture with a bar shape extended onto three columns. Similar experiences can be drawn from other form-finders as well. While looking through the devices, the viewers get a deeper sense of the site.

Why do they want to find Malevich? His suprematism theory is the product of an industrializing Russia, and the silo is a classic of industry structures. The installments are intended to lead viewers implicitly through a journey on the relationship of art and industrial aesthetics as they try to find Malevichian geometric figures.

T5 / FANG Zhenning/ Cohesion
By using the rest of the space of special exhibitions in the 257 Factory, this work facilitates to change immovable indoor facilities, otherwise obstacles, into a piece of large spatial installment composed of various three-dimensional geometric objects, collectively known as "cohesion". The biggest object is a 5-meter cube, enlarged from a 4.8×4.8×3 meter storage space. Its prototype is Red Square, created by Malevich in 1915. Cohesion is placed around the Malevich Visual Chronology as a hint to its connection with suprematism. Another 5-meter black square is placed on the side that was designated to block a large sliding door of similar size, but I can't resist the temptation to turn it into something aesthetically meaningful.

We use geometric shapes as language symbols to accentuate the space, to recall the climax of cubism as an analytical process. The concept of "Cohesion" is a contemporary version to show respect to suprematism, composed of a red irregular cube, a triangular prism, a black cross and a black square.

P10 / FANG Zhenning/ Malevich Visual Chronology
Kazimir Malevich (1879-1935), a major artist in the 20th century and theorist of Suprematism, has a great influence on the international art community in the last century, shown only after contemporary art was thoroughly studied. Why does Malevich enjoy such great influence among so many Avant-Garde artists in the same period? His works are considered both Russian and worldly treasures because of their innovation and spirituality, fostering a unique symbolism that has been inherited and developed up to the present. American Minimalist artists were influenced by his creative efforts half a century later. The editor has been collecting and researching information on Malevich for about thirty years, and this visual chronology is based on the results, or "big data". This chronology is a revised version of "Malevich Documenta", which was the first systematic summary and analysis of Malevich's suprematist works in Chinese academic community, organized for the one hundredth anniversary of Malevich's Suprematism Statement in 2014. It's safe to say that the overall spatial arrangement of the 257 Factory is a suprematist experiment.

P11 / FANG Zhenning/ Wanxiang
"Wanxiang" is a gallery of paper sheets and a consumption-oriented interactive project that has received success beyond imagination, providing about 20 artists, architects and publishers an opportunity to interact with the audience. Eighteen posters are settled and arranged linearly on a 30-meter space. The interaction is straightforward: each poster is printed in full size, or 635×965cm, and in 10,000 copies. That is, 180,000 posters are spread at the far end of the exhibition space so that visitors must go through the entire hall before taking away posters as they wish. We consider it as a unique way of communication and spreading information, an act of behavioral art. Two weeks after the launch ceremony, half the stock of the posters were taken away by visitors. After six weeks, not a single copy remained.

Exhibition is Installment

The main exhibition of SUSAS was launched in two venues: the 80,000-ton silo and the 257 Factory. In terms of size and capacity, the 257 Factory area covers more than 3,000 square meters, not necessarily a competitor to the silo which stands over 50 meters above ground, and covers an area more than 4,000 square meters. Yet it has a relatively uniform layout with an available height of 11 meters, posing a difficult problem for curating in this spacious room. I believe that curation and plans need to be strategically devised in relation to concept and space manipulation so that the original look, exterior or interior, of the industrial building should be kept intact, especially when such a valuable heritage is concerned, thus we tried not to block the existing structure. For example, we would follow the construction framework and let it inspire us on how to arrange the exhibition. "Place Dictates Form" still holds as a principle that we have carried through in planning the exhibition of the 257 Factory.

These are the five elements of the set-up:

1) Exhibition design is a highly functional and creative attempt that requires an organic integration of content and form. However, I wish to avoid all the common pitfalls that have produced many mundane works in this sector. What I want is an exhibition that is par excellence, an aesthetically intriguing installment.

2) No view-blocking and heavy walls are included in this plan, and light fabrics are used to hang exhibition items, much in the same way we hang clothes. This act makes set-up and dismantling convenient, and follows the ecological principle that I have been advocating all along, that is, zero garbage in setting up the exhibition.

3) Hang all the exhibition panels in the air, so that viewers may walk around more freely.

4) Existing industrial facilities are transformed into hardware available for the exhibition. For instance, a small storage space is turned into a cubic installment, and iron ladders become a platform for showing picture albums.

5) The addition of hung items is a risky act because our limited experience doesn't offer an accurate estimate about the thickness of square poles necessary to hold the items. We designed four 3×3 grids with the size of 9×9 meters, each for one special exhibition. The length of poles is 8 meters, and to minimize visual bulkiness and guarantee solidity, the cross-sectional size is 2.8×2.8 cm (vertical poles) and 3.2×3.2 cm (horizontal). A horizontal pole is attached above each thin vertical pole for structural strength. Hooks are not applied to maintain intactness and perfection of the grid design. The top of the hanger is 11 meters above ground, while the height of each exhibition item is kept under 1.2m. The nearly 90% of unused height highlights the remaining functional area, making the seemingly useless space more purposeful.

Though I have been curating for ten years, SUSAS 2017 is my first experience with large urban public spaces. Based on approaches to infiltrate art into spaces and activate silent corners, we found a potentially effective way to renew and regenerate cities that has aroused significant public reactions. Doubtlessly, SUSAS would be a reference for any event of the same type in other cities. However, the success of one case can never be easily copied. Each exhibition faces new problems, and its curation always segues into a new adventure. What remains sound for art interventions in urban public spaces is the principle "Place Dictates Form", that is, how an art piece is created for a public space should be determined by the quality and nature of its site.

时间与地点的再定义
REDEFINING TIME AND PLACE
民生码头八万吨筒仓建筑的临时性改造与再利用
TEMPORAL REPURPOSING AND REUSE OF THE 80,000-TON SILO AT MINSHENG PORT

柳亦春 大舍建筑设计事务所主持建筑师
LIU Yichun Presiding architect of Atelier Deshaus

八万吨筒仓是民生码头中最具震撼力的工业遗产，虽然建成时间只有短短的22年，却作为不会再出现的建筑空间类型而具有历史遗产的保护价值。按照著名艺术史学家阿罗伊斯·李格尔的分类法，它属于"非有意创造的纪念物"。作为曾经的生产建筑，其原本的生产功能在城市的发展进程中逐渐褪去，留下空却的建构物已如废墟般存在，这时曾经在这个空间中所发生过的劳作不再成为关注的焦点，反而是作为废墟的筒仓在建造时其建造逻辑因为背后的工业生产的工具理性而突然成为城市中的野性力量，令人赞叹不已。

对待工业遗产，"更新"的观念与"原真性"保护修缮理念似乎永远存在某种矛盾，而事实上原真性在建筑脱离其原本的时代和社会背景的条件下也是不可再现的，工业遗产的修缮和保护更应在延续和保护其历史价值和文化意义的基础上，使其在新的时代和社会背景中获得新的价值和意义。

由国际古迹遗址理事会（ICOMOS）澳大利亚国家委员会所制定的《巴拉宪章》为文物建筑寻找"改造性再利用"的方式越来越受到重视，并在工业遗产保护项目上加以推广。"'改造性再利用'关键在于为某一建筑遗产找到恰当的用途，这些用途使该场所的重要性得以最大限度地保存和再现，对重要结构的改变降低到最低限度并且使这种改变可以得到复原。"《巴拉宪章》所定义的"改造性再利用"指的是对某一场所进行调整使其容纳新的功能，这种做法因没有从实质上削弱场所的文化意义而受到鼓励推广，"寻找恰当的用途"应当成为工业建筑改造一个非常重要的前置性条件。

八万吨筒仓作为2017上海城市空间艺术季的主展馆，便是在这一"改造性再利用"的原则下所进行的一次空间再利用的积极尝试，以艺术展览为主要功能的城市公共文化空间是为八万吨筒仓所寻找的非常适合的功能，能最大程度地符合现有筒仓建筑相对封闭的空间状态。

本次艺术季主展馆主要使用筒仓建筑的底层和顶层，由于筒仓建筑高达48米，要将底层和最顶层的空间整合为同时使用的展览空间，必须组织好顺畅的展览流线，同时也要处理好必要的消防疏散等设施。本次展览流线组织的最重要的一个改造动作是通过外挂一组自动扶梯，将三层的人流直接引至顶层展厅。这样人们在参展的同时也能欣赏到北侧黄浦江以及整个民生码头的壮丽景观，除了悬浮在筒仓外的外挂扶梯，筒仓本身几乎不做任何改动，极大地保留了筒仓的原本风貌，同时又能看到重新利用所注入的新能量。这个改造动作直接面对了筒仓改造的主要矛盾，即原本封闭的仓储建筑在转为公共文化空间时如何获取必要的开放性？如何建立在新时期的时间性与场所感？这组外挂扶梯无疑重新定位了八万吨筒仓的位置：通过引入浦江景色去揭示它坐落在黄浦江边这一事实，同时将滨江公共空间带入这座建筑。建筑的公共性由此获得，一种新的时间也被铭刻在旧有的时间上。

在外挂扶梯的底部，我们还与艺术家展望合作，利用艺术家独特的拓片肌理的反射不锈钢板作为外挂扶梯的底面装饰，它倒映着民生码头周遭的景象，而外挂的这组巨大的扶梯体量也因此变得轻盈。未来，随着从江边直上筒仓三层的粮食传送带被改造为自动人行坡道，一个从江边可以直接上至筒仓顶层的公共空间得以建立，这个壮观的公共空间将成为浦东滨江贯通和民生码头空间更新项目之间的重要纽带。这里将成为一个带着上海工业文明历史底蕴、以艺术结合城市功能并积极介入市民日常生活的公共艺术空间。

The 80,000-ton Silo is the most striking industrial heritage site at Minsheng Port. The silo is worth careful preservation, despite its relatively short 22-year life, representative of a type of structure with no active decedents in the future. According to art historian Alois Riegl, it should be categorized as an "Ungewolte Denkmal", or an unintended monument. The productive function designated as it was built, together with workers and laborers, have dropped out of place with a developing Shanghai, leaving an empty structure without fuction, whose underlying instrumental reason crystalized in the industrial logic and process of its construction has made the place a wild focal point of this city in a short time period.

"Renewal" and "Authenticity" pose a dilemma for anyone concerned with renovating industry heritage. Authenticity, at its face value, is an unachievable goal since the temporal and social context that used to bestow meaning to the physical structure is already gone. I believe a renovated heritage should receive new values and meaning that makes natural sense in a changed context, based on the preservation of its historical values and cultural symbolization.

"Adaptive Reuse", developed by Australia ICOMOS (International Council on Monuments and Sites), is an approach increasingly appreciated and implemented on industrial heritage projects. The key of adaptive reuse is to find a restorable use that maximally preserves and reproduces the significance of the structure, and minimally changes the fabric. "Adaptive Reuse" refers to a transformative process of a site that enables new functions and doesn't damage its cultural significance in any way, making it a process worth knowing and applauding. "Find a use" should be a precondition of any renovation project on industrial heritage.

As the main exhibition space of SUSAS 2017, the 80,000-ton Silo is a proactive experiment of repurposing space, guided by the "adaptive reuse" principle. An urban public cultural space, mostly for art exhibitions, makes perfect use of the silo for its enclosed space.

Two floors, ground and top, of this 48-meter high building are used as the main exhibition halls, which means that well-designed visitor flow, fire safety and evacuation are indispensable if they are to be integrated into a whole. The most significant adaption for flow organization is an escalator attached to the silo's exterior that transports visitors at the third floor to the top, with a view of the Huangpu River in the north and the entire Minsheng Port. Few changes are involved, except for the escalator, keeping the silo's historic integrity. Yet new energies are noticeable from the reuse project. The transformative action confronts the chief dilemma of this project: how can an enclosed warehouse be turned into a cultural public space, which must be open? Furthermore, how to establish a new time and place? The attached escalator repositions the 80,000-ton silo by introducing a river view into the picture, and therefore makes explicit the fact that it is situated near the Huangpu River. The other element introduced simultaneously is the public waterfront space. We now have a genuinely public structure with new temporal layers over its past.

Decorations specifically created by artists are at the bottom of the escalator. These reflective stainless-steel sheets with unique rubbing patterns not only show the reversed image of Minsheng Port and its neighborhood, but lightens the experience with the huge escalator itself. In the future, the conveyor that used to lift grain from the bank to the silo's third floor will be transformed into a moving slope for visitors, finishing a bank-to-top public space. It will be a nexus of the Open Space Leading Down the East Bund of the Huangpu River and the Minsheng Port Renewal, as well as a public art space carrying Shanghai's industrial past, integrating art into urbanity and with daily life.

项目名称：民生码头8万吨筒仓改造项目——2017上海城市空间艺术季主展馆
建筑师：大舍建筑设计事务所
合作设计：同济大学建筑设计研究院（集团）有限公司
建筑设计小组：柳亦春、陈晓艺、王伟实、王龙海、张晓琪
项目地点：上海市民生路3号
项目功能：临时展览
设计时间：2016 – 2017
建成时间：2017.10
改造建筑面积：约16320平方米
业主：上海东岸（集团）有限公司
施工单位：上海一建集团有限公司

Project: Renovation of 80,000-ton Silo at Minsheng Port — Main Exhibition Hall of Shanghai Urban Space Art Season (SUSAS) 2017
Architect: Atelier Deshaus
Structure/M&E: Tongji University Design (Group) Co., Ltd
Design Team: LIU Yichun, CHEN Xiaoyi, WANG Weishi, WANG Longhai, ZHANG Xiaoqi
Location: No.3, Minsheng Road, Shanghai
Program: Temporal Exhibition
Design period: 2016 - 2017
Completion: October 2017
Floor Area: c. 16,320 m²
Client: Shanghai East Bund Investment (Group) Co., Ltd
Construction: Shanghai Construction No. 1 (Group) Co., Ltd

主题 thisCONNECTION 阐释：连接的多义性
AN INTERPRETATION OF "thisCONNECTION": POLYSEMY OF CONNECTION

thisCONNECTION\disCONNECTION，读音上非常接近，却是意义相反的两组词汇。我们借用这两组对立意义的词汇来暗喻我们的公共空间生产所呈现出来的状态。在公共空间的生产、人与人关系的重组以及秩序建立的过程中，不恰当的设计或规划带来了更加深层的断裂。南非的城市街道，一边是富人区，一边是饥寒交迫的贫民窟。美国墨西哥边境的边境线，深深的高墙却预留了视线穿透的孔洞，每个周末在美国打工的墨西哥人民跑到这里与家人"见面"——隔着高墙的见面是手指的触摸和声音的传递。这些情景能够让我们感受到断裂与连接之间的动态转化的能量。如何将这种藕断丝连的修复力量挖掘出来并使之强化，从而来生产/修复我们公共空间中的断裂状态，正是这次展览所要探讨的主题——thisCONNECTION 连接：连接不是简单的线程之间的连接，我们探讨的连接包含空间中身体的连接，空间中时间的连接，空间中物理的连接，空间中的社会连接，以及空间中的文化连接。

空间中的身体连接：如果我们的身体可以丈量空间，那么步行是最好的丈量方式。在步行中，我们的身体在观察，在记录，在体验以及在与其他的身体发生碰触、故事以及记忆。随着社会的发展，步行空间在褪去，空间也因为速度而变得更加模糊与相似。这种模糊带来的断裂，是空间面貌唯一性的逝去。置于展场中的装置《内省腔》，来自艺术家尹秀珍，用那些携带旧主人气息的衣物/纺织物重新编织，唤起人们在母体子宫内的记忆。她说："人类最初的状态就是在母体的羊水中，我希望人们能回归这种状态，进行自我观照。因为在现代社会中，人很容易迷失，失去判断，变得茫然无从，不知道自己真正需要的是什么。"互动性很强的腔体内，一个个独特而鲜活的社会标本进入探索，从而也在个体之间形成全新连接的社会关系，创造记忆性的身体连接。

空间中物理的连接：空间自然的割裂一直是人类努力克服的天堑。而这种技术的超越所形成的连接有着直观而强大的力量。这种直观的"连接"也以丰富的形式出现：技术、结构与新材料在这种直接的连接形式里扮演了举足轻重的作用。而简单的连接之外也带来了很多意想不到的结果：纽约 Highline 高线公园由跨越 23 个街区的高架铁路桥改造而来，丰富了城市界面，更融合了纽约这座城市不同的环境与记忆；吉首美术馆由上下两座桥叠合而成，像条街道的钢桥与包容画廊的混凝土桥直接围合出了美术馆的展厅，这些展览新颖地探讨了作为物理的桥的复合功能形式，也承载了文化发生的场所；当然简单的技术革新也给我们通向未来带来了更多可能性，包括 hyperloop 给我们展示的技术带来的未来憧憬。

空间中的时间连接：梅洛·庞蒂在皮埃尔·雅内的《记忆进程和时间概念》中阐释，"空间和存在性这两者叙始终不断地相互定义、相互影响着，它们的集合是我们与世界的关系——体验。"影像艺术家多米尼克·冈萨雷斯-福斯特用城市影像将时间定格，蒙太奇的叙述有关城市影像和记忆。装置艺术家用特定时空中的材料将思考定格，创造记忆的穿梭与重现。在筒仓序列空间内这些艺术家用自己的语言穿越不同时空片段，形成连接。

空间中的社会连接：空间是社会活动的载体，在承载社会性的同时，因为本身迅速而过度的膨胀所带来的各种出乎意料的屏障，形成了社会群体之间的交流障碍以及对立。如何创造真正的公平空间，使社会的各个层面都能共享空间，为社会各个层面提供真正的庇护，支持人类面对无边的都市，也成为我们这次展览讨论的重要课题。Placido Gonzalez Martinez 的《回音》就借用建筑、多媒体、图像等手段来反映社会问题与空间生产的联系。

本次展览的发生地——民生码头作为城市滨水公共空间的一部分，本身以具体改造案例的身份去呈现一个断裂城市空间的转化可能，从构筑新的开放的滨水平台开始，以真正提升黄浦江两岸开放空间的潜在价值，从而以"thisCONNECTION"的姿态重新出现，并以此促成更多有着相似"连接"性的未来公共空间。而我们所探讨的"连接"，正是帮助民生码头完成从"闲置码头"到"未来公共空间"的身份转换的能量与媒介。

Despite phonetic similarity, thisCONNECTION and disCONNECTION constitute a semantically polarized pair, which we use as a metaphor for the situation produced by our public spaces. As public spaces are manufactured, interpersonal bonds are reorganized and new orders are established, with deeper fractures caused by inappropriate design or planning. Behold the South African streets. Here is a wealthy district, and on the other side is a ragged slum. Holes are left on the Mexico-United States barrier so that Mexicans who work in America can see their families on weekends. All these scenarios impose on us a new energy inspired between the struggle of connection and facture.

That's what the theme of SUSAS 2017 is about: thisCONNECTION. It's about unearthing and reinforcing the tacit power of renovation that produces or renews our fractured public spaces. Here, we aren't connecting dots or lines, but making bodily, temporal, physical, social and cultural connections in spaces.

Bodily Connection: The best bodily way to measure spaces, if any, is walking. As we walk, our bodies are natural observers. Collisions, stories, memories are created as we interact with other bodies. Civilizational development propels walking spaces into background and corners, and with its modern speed, assimilates and blurs all spaces. Spaces used to be unique, yet no more because of blurring fractures. "Introspective Cavity" is an exhibited installment created by Xiuzhen Yin, a remix of collected clothes or fabrics with scents of their former owners that aims to arouse the viewer's memory in utero. As the artist said, "the uterus is the starting point of our lives. The hope is that viewers can return to this state for self-reflection and self-examination. People easily get lost, losing judgment in modern society. They don't know what they really need." In this highly interactive cavity, a series of unique specimen of our society is probed. A viewer can establish otherwise non-existent social bonds with other individuals, and create a memorial connection with his or her body.

Physical Connection: Humans have been trying to overcome natural fractures, all these resulting in artificial connections that embody forces immediately felt and come in a variety of forms thanks to technology, structure and new materials. Surprising results are many beyond connection itself. For instance, Highline Park in New York is renovated from an elevated rail through 23 blocks that furnishes interfaces to the city and blends in its unique context and memory. Jishou Art Museum constitutes a street-like steel bridge and a concrete bridge above it that houses galleries, with all the exhibition spaces situated in between. The museum explores the multi-functional possibility of a bridge as a physical construction, and is a place where cultural activities happen. Technological innovations may also open the way to future opportunities, like how the hyperloop foreshadows tomorrow's hopes brought about by technology.

Temporal Connection: Maurice Merleau-Ponty explained in the book Process of Memory and Concept of Time written by Pierre Janet that "Space" and "Existence" define and influence each other in the constant. Their intersection is our relationship with the world - Experience. Dominique Gonzalez-Foerste eternalizes moments in cities with his photos, offering a montage narrative about urban images and memory, while various installment artists eternalize ideas with materials from specific spatial-temporal locations as a way to observe and revitalize what awaits to be remembered. In this orderly space of silos, they have travelled through a myriad of segments in time and space, using their respective language as a medium. In this way connections are made.

Social Connection: Space houses our social life, yet as it swells disproportionately, it unintentionally cuts off communications, blocks communities and promotes antagonism. How to create a truly equalized space for every group of society to share? How to truly shelter every member of society? How to support people faced with many metropolitans? These are all major questions on our issue list. Authored by Placido Gonzalez Martinez, "The Echo from Society" is an attempt to reflect the relationship between social problems and space creation utilizing architecture, multimedia and other imagery.

Minsheng Port, the venue of this exhibition, is one of the city's waterfront public spaces and a concrete instance of how renovation presents the transformational possibility of a fractured city. The renovation project began with building a new public waterfront platform in order to promote the potential values of public spaces along both banks, and stands as a realization of "thisCONNECTION" that facilitates more public spaces with similar "connectivity" in the future. "CONNECTION", as we discuss here, is what empowers and mediates the amazing transformation of Minsheng Port from an "idle dock" to a "future public space".

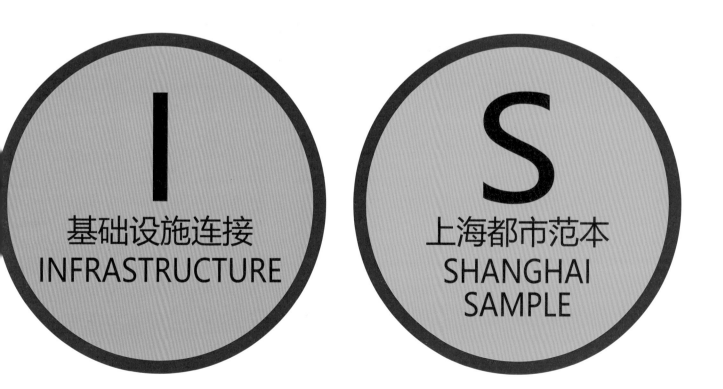

户外展区

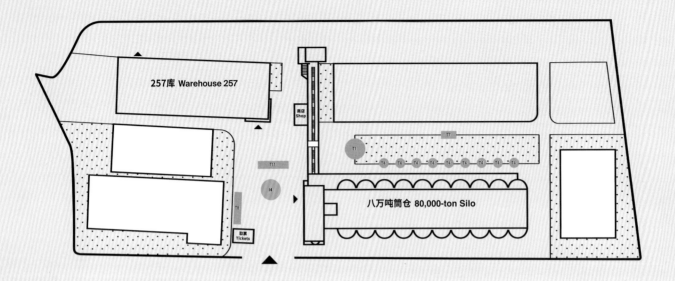

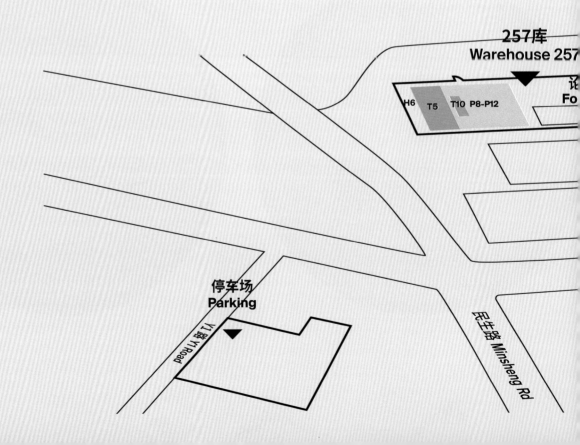

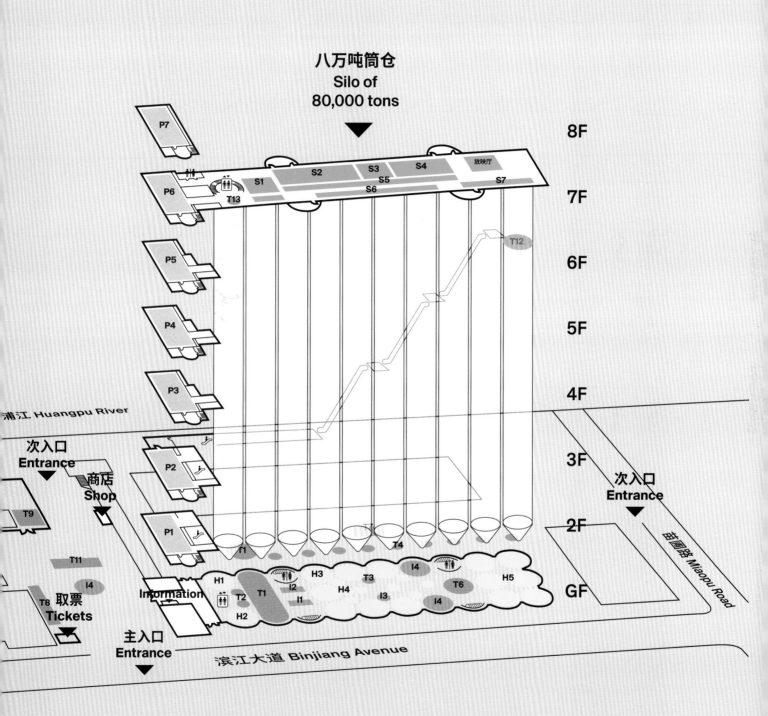

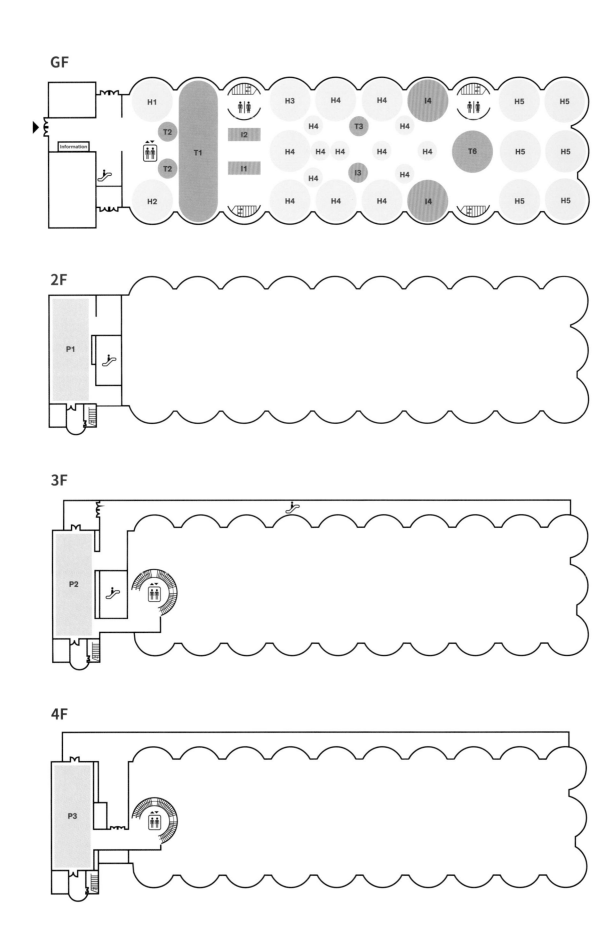

5F

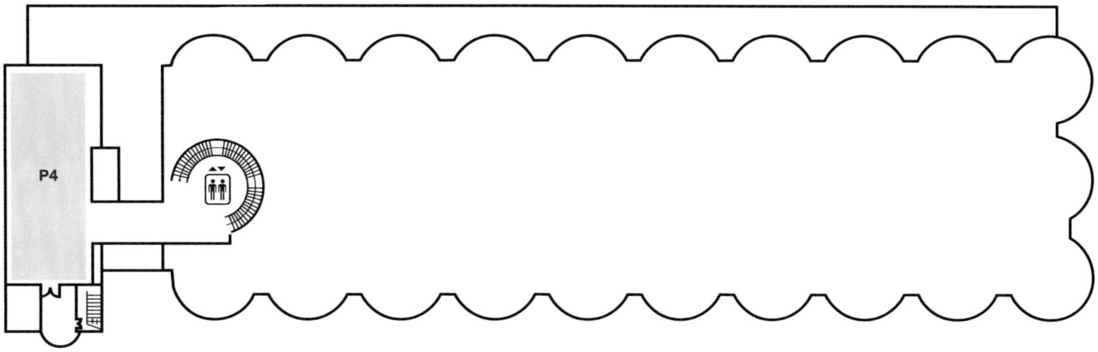
P4

6F

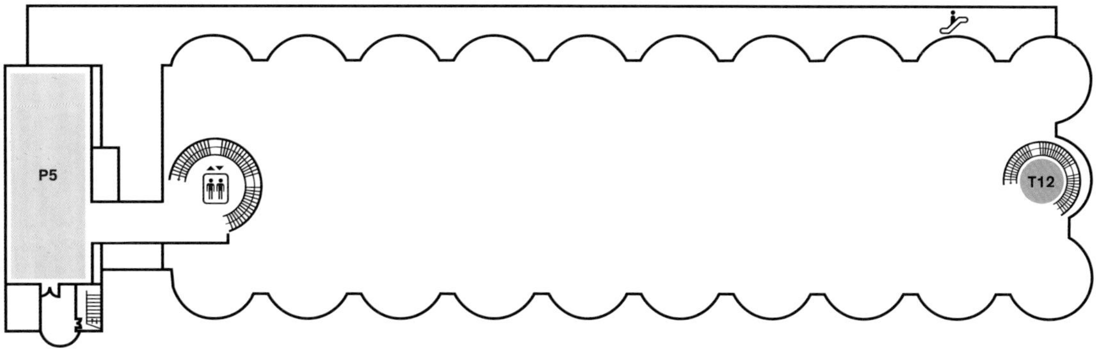
P5
T12

7F

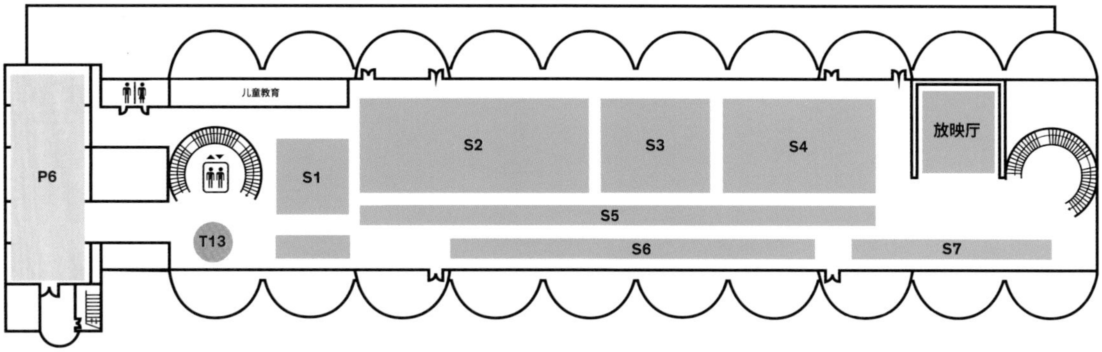
儿童教育
P6
S1
S2
S3
S4
放映厅
S5
T13
S6
S7

8F

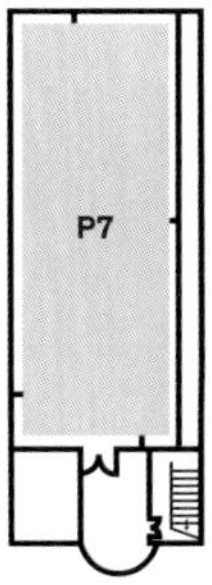
P7

257

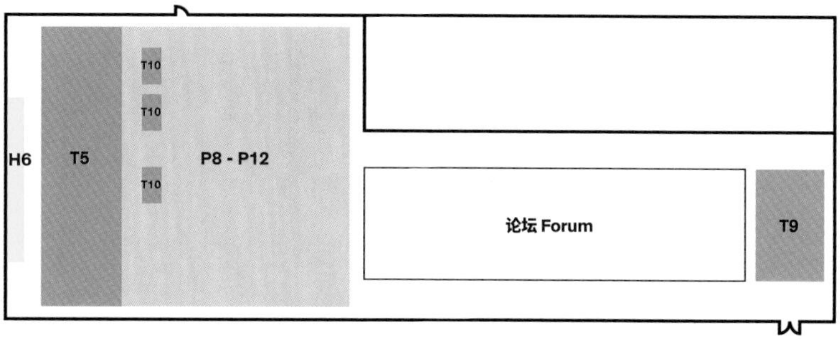
T10
T10
H6
T5
P8 - P12
T10
论坛 Forum
T9

特 SPECIAL PAVILION 展

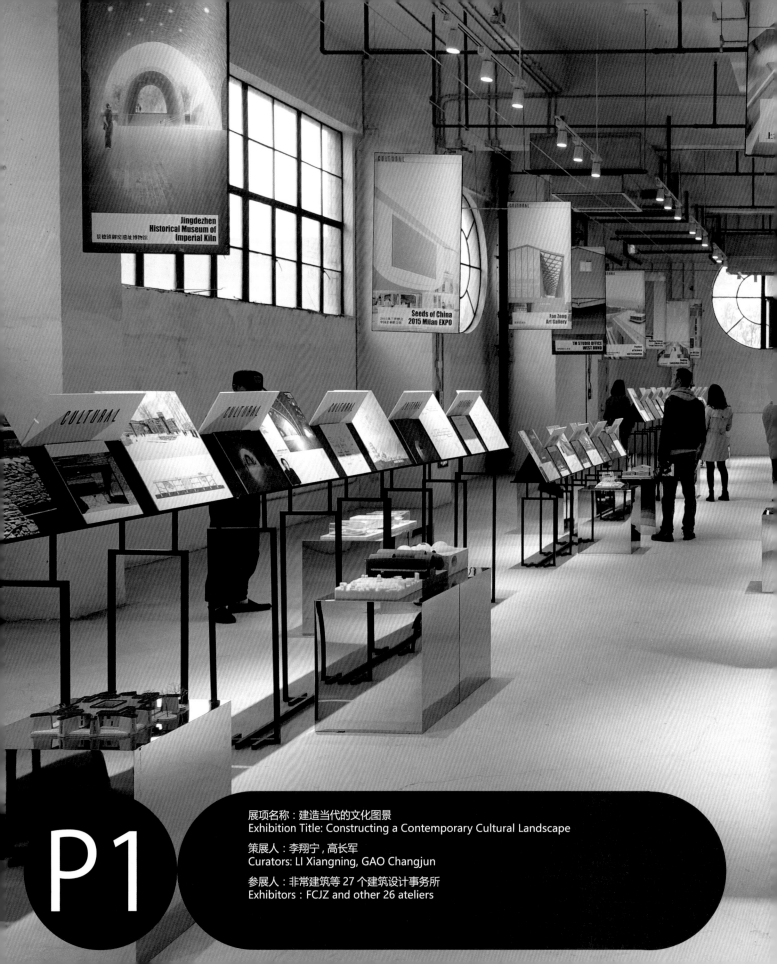

展项名称：建造当代的文化图景
Exhibition Title: Constructing a Contemporary Cultural Landscape

策展人：李翔宁，高长军
Curators: LI Xiangning, GAO Changjun

参展人：非常建筑等 27 个建筑设计事务所
Exhibitors: FCJZ and other 26 ateliers

新兴的当代文化建筑基于对时代的批评阅读，力图建立历史遗产与现代文明、精神的纯创造与繁盛的物质生产等之间的连接。在"祛魅"与"返魅"、自主性与群像化的思想张力中，当代建筑实践在文化创造与文化批评等领域中催生出千姿百态的策略。本展位呈现近三十个当代中国近年来最具代表性的文化建筑作品，涵盖博物馆、展览馆、美术馆、市民活动中心、办公创意园区等类型。

通过荟萃文化建筑自身的作品性与其所含内容的多元性，建筑师探索了空间与人文交融的广阔前景，以新时代性构筑新的文化图景。在新一轮的文化建构进程中，实践的多元策略与迂回路径、理论的独立立场与批判角色均得到更深层次的研究与展现。

Based on a critical review of the current era, emerging cultural architecture promotes itself in connecting historical heritage and modern civilization, of the pure creation of spirit and redundant mass production. Drenched in the conflicts of disenchantment and Re-enchantment, autonomy and mainstreaming, contemporary practice in architecture has developed various strategies in both cultural production and criticism. With about 30 projects coming from all over the country, including museums, exhibition halls, galleries, public activity centers, and creative parks, this exhibition is meant to present the vast potential of interaction between space and humanity.

Selected projects serve as a specific review of a prosperous contemporary cultural landscape, in which further studies are conducted and expressed, with resilient response time, detoured trajectory in practice and independence, and critical role in theory.

A01 张永和 非常建筑 上海诺华制药／园区实验楼
A01 Novartis Shanghai Campus Laboratory Building / Yung Ho Chang, Atelier FCJZ

A02 崔恺 崔恺工作室 玉树康巴艺术中心
A02 Yushu Khamba Arts Center / CUI Kai, CUI Kai Studio

A03 戴璞 大章建筑 树美术馆
A03 Tree Art Museum / DAI Pu, DAI Pu Architects

A04 董功 直向建筑 三联海边图书馆
A04 Seashore Library / DONG Gong, Vector Architects

A05 傅筱 青岛世界园艺博览会梦幻科技馆
A05 Pavilion of Science and Technology at the International Horticultural Exposition 2014 Qingdao / FU Xiao

A06 韩涛 Than-Lab 工作室 中国油画院
A06 Chinese Academy of Oil Painting / HAN Tao, ThanLab Office

A07 华黎 迹·建筑事务所 三里屯 Tiens Tiens 甜品店
A07 Tiens Tiens / HUA Li, Trace Architecture Office

A08 李立 同济大学建筑设计研究院（集团）有限公司 洛阳博物馆
A08 Luoyang Museum / LI Li, Tongji Architectural Design and Research Institute

A09 李麟学 麟和建筑工作室 南开大学津南校区活动中心
A09 Jinnan Campus Student Center of Nankai University / LI Linxue, Atelier L+

A10 李兴钢 李兴钢工作室 元上都遗址工作站
A10 Entrance for Site of Xanadu / LI Xinggang, LI Xinggang Studio

A11 刘克成 刘克成工作室 中国科举博物馆
A11 Chinese Imperial Examinations Museum / LIU Kecheng, LIU Kecheng Studio

A12 柳亦春 + 陈屹峰 大舍建筑设计事务所 龙美术馆（西岸馆）
A12 Long Museum West Bund / LIU Yichun+ CHEN Yifeng, Atelier Deshaus

A13 陆轶辰 清华大学、Link-Arc 建筑师事务所 2015 米兰世博会中国馆
A13 China Pavilion for Expo Milano 2015 / LU Yichen, Tsinghua University+Studio Link-Arc

A14 任力之 同济大学建筑设计研究院（集团）有限公司 "中国种子"——2015 米兰世博会中国企业联合馆
A14 'Chinese seed' — China Corporate United Pavilion of Expo 2015 Milan / REN Lizhi, Tongji Architectural Design and Research Institute

A15 阮昊 零壹城市 天台县赤城街道第二小学
A15 Tiantai No.2 Primary School / RUAN Hao, LYCS Architecture

A16 童明 童明建筑工作室 西岸工作室
A16 TM Studio [West Bund Office] / TONG Ming, TM Studio

A17 王路 清华大学建筑学院王路工作室 / 壹方建筑 毛坪村浙商希望小学
A17 The ZS Hope Primary School / WANG Lu, School of Architecture, Tsinghua University / In+of

A18 王维仁 王维仁建筑设计研究室 西溪湿地艺术村 N 地块
A18 Xixi Wetland Artist Village (N site) / WANG Weijen, WANG Weijen Architecture

A19 王昀 方体空间工作室 西溪学社
A19 Xixi Learning Community / WANG Yun, Atelier Fronti

A20 魏春雨 地方营造工作室 张家界博物馆
A20 Zhangjiajie Museum / WEI Chunyu, WEI Chunyu Studio of Construct

A21 吴钢 维思平建筑设计 北京龙山教堂
A21 Beijing Longshan Church / WU Gang, WSP ARCHITECTS

A22 张斌 + 周蔚 致正建筑工作室 同济大学浙江学院图书馆
A22 Zhejiang Campus Library of Tongji University / ZHANG Bin+ZHOU Wei, Atelier Z+

A23 张利 简盟工作室 嘉那嘛呢游客到访中心
A23 Jianamani Visitor Center / ZHANG Li, Atelier TeamMinus

A24 章明 + 张姿 原作设计工作室 范曾艺术馆
A24 Fan Zeng Art Gallery / Zhang Ming+Zhang Zi, Original Design Studio

A25 朱竞翔 Unitinno 肯尼亚 MCEDO 贫民窟学校扩建
A25 Kenya MCEDO School Expansion / ZHU Jingxiang, Unitinno

A26 朱锫 朱锫建筑事务所 景德镇御窑遗址博物馆
A26 Jingdezhen Imperial Kiln Museum / ZHU Pei, Studio ZHU-Pei

A27 祝晓峰 山水秀建筑事务所 华鑫中心
A27 Huaxin Center Shanghai Google Developers Group Starup Incubator / ZHU Xiaofeng, Scenic Architecture Office

METROPOLIS

NEW ISSUES FOR THE FUTURE OF THE CITY:
BARCELONA METROPOLIS OF CITIES

城市未来新论：
巴塞罗那大都市圈

本展览主题是巴塞罗那大都市区域及其近期的转型变化。展览最初由巴塞罗那大都会政府组织，于 2015 年在巴塞罗那第一次展出，并于 2016 年在哈佛大学设计学院改版再次展出。

《巴塞罗那：城市群大都会》展示了自 20 世纪 80 年代以来巴塞罗那地区一系列重要且具创新性的城市发展项目，也从新角度解读了其他欧洲城市的发展项目，展示了全球不同城市应对当代城市问题的大型项目和策略范例，并开启了有关 21 世纪的生态、经济、社会性对于塑造新城市空间形态影响的讨论，探讨在不同文化背景下环境、能源、生态交通、社会公平及其他要素如何影响新的城市策略，从而帮助城市更高效地运转。

This exhibition hails from the Barcelona metropolitan territory, showing its recent transformations. It was organized by Àrea Metropolitana de Barcelona and presented in Barcelona in early 2015 and received a new treatment at the Graduate School of Design, Harvard University in fall 2016.

Barcelona: Metropolis of Cities presents a series of critical urban developments since the 1980s that created new ambition for Barcelona and the metropolitan cities surrounding it. At the same time it cast new light on Europe's current developments and showcases urban projects and strategies that engage contemporary challenges cities have to face. The Exhibition opens up the debate to highlight the extent to which twenty-first-century ecological, economic, and social issues are giving shape to new urban forms. Concerns for the environment, energy, sustainable mobility, and social equity (among others) characterize new urban strategies in many different contexts around the world that enable cities to function more efficiently and effectively.

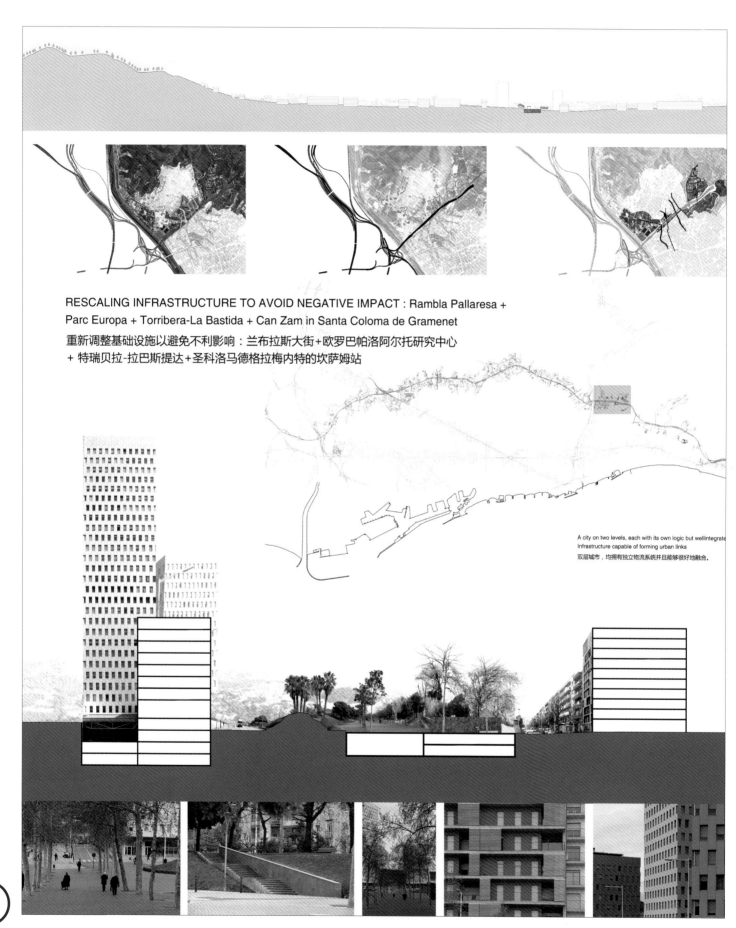

RESCALING INFRASTRUCTURE TO AVOID NEGATIVE IMPACT : Rambla Pallaresa + Parc Europa + Torribera-La Bastida + Can Zam in Santa Coloma de Gramenet

重新调整基础设施以避免不利影响：兰布拉斯大街+欧罗巴帕洛阿尔托研究中心 + 特瑞贝拉-拉巴斯提达+圣科洛马德格拉梅内特的坎萨姆站

A city on two levels, each with its own logic but well integrated. Infrastructure capable of forming urban links

双层城市，均拥有独立物流系统并且能够很好地融合。

展项名称：拉斯维加斯工作室：来自罗伯特·文丘里和丹妮丝·斯科特·布朗档案馆的影像集
Exhibition Title: Las Vegas Studio: Images from the Archives of Robert Venturi and Denise Scott Brown

策展人：希拉·斯塔德勒，马提诺·斯泰尔利
Curators: Hilar Stadler, Martino Stierli

参展人：比尔帕克博物馆
Exhibitors : Museum im Bellpark

发表于1972年的专著《向拉斯维加斯学习》带领我们领略了罗伯特·文丘里和丹妮丝·斯科特·布朗1968年在耶鲁的研究和设计工作室中引人入胜的可视化对谈。我们将图片从他们最初的分析文本中提取出来，并且将它们以影像感知的形式加以展出。我们的企划则将重点拉回至他们理论形成之前对于影像资料本身的关注，我们的选择很大程度上聚焦于他们研究方案中的次要方面以及旁生结果，同时也不忽略那些经典的影像。展览因此将先前不为人知的图像带到幕前。我们相信，文丘里和斯科特·布朗在拉斯维加斯调查中的真正兴趣在这些"无意识"瞬间当中得到了体现。

拉斯维加斯工作室创造了建筑摄影、城市研究、理论和本地贸易之间的关系。该项目所采用的研究手法又与上海当代城市研究产生了共鸣，使得一场跨越时空的生动对谈得以实现。

Learning from Las Vegas, a treatise on architectural theory published in 1972, captivates us primarily through its engaging visual discourse developed during Robert Venturi and Denise Scott Brown´s 1968 research and design studio course at Yale. We removed the pictures from their original analytical context and present them in photography. Our project returns to a point before theory formation and refers directly to photography itself. Our selection focuses largely on secondary aspects and side products of the research project without losing sight of the iconic imagery. It shifts to the previously unknown images. We believe that the true interest in Venturi´s and Scott Brown´s approach to Las Vegas becomes clear in these "unconscious" moments.

"Las Vegas Studio" established a relationship between architectural photography, urbanism, theory and the commercial vernacular. The methodology in this project resonates with Shanghai's contemporary urbanism, allowing for a meaningful dialogue across space and time.

P4

展项名称：液态历史：泰晤士河的想象与现实
Exhibition Title: Liquid Histories: the Thames, between the Real and Imaginary

策展人：大卫·尚贝尔，杰里米·蒂尔，马丁·维尔
Curators: David Chambers, Jeremy Till, Martyn Ware

参展人：中央圣马丁艺术与设计学院
Exhibitors: Central Saint Martins

"液态历史"指的是通过多种观点和方法来讲述历史，借历史的流动性来冲刷事实的固定性。泰晤士河的历史就像液态的水流——多样、重叠，从独立的个人叙述到伟大的国家历史，从诗人的神话故事到房地产公司的虚假事迹都囊括其中。

作品通过探索泰晤士河想象与现实间的张力，展示了泰晤士河液态的历史。它反映了这条河流想象中本为公共空间，而现实中的法规和经济发展却与之背道而驰。这条长椅模拟了泰晤士河从黑衣修士到达格南一带的路线和形状。在装置背面，铭牌上一系列的引述呈现了泰晤士河的真实情况，当中包括泰晤士河逐渐发展成为新的住宅地段的过程、因此导致的经济后果，以及两岸原居民被驱逐在外的情况。

由世界著名的声音艺术家和音乐家——马丁·维尔设计的三维音景，让游客的想象力沉浸在泰晤士河的声音、故事和感觉中。在想象与现实这两极之间，一系列影片更能呈现泰晤士河作为一个日常生活中的媒介空间的强大力量。

作品展示泰晤士河多重景观的目的在于表现任何地方的河川都必须保存其隐秘、寂静的形态，而不能因其如画景观而被开发成商品。

The title of the installation refers to how history can be told through multiple viewpoints, lending history a fluidity to wash away the fixity of a list of facts. The histories of the Thames are diverse and overlapping like liquid. From individual and personal narratives to the grand histories of nationhood, with everything from the mythical histories of poets to the false histories of estate agents in between.

The installation presents these liquid histories of the Thames by exploring the tension between what is imaginary as opposed to reality. More specifically, the river postulated as a public space and the frustrated actuality for reasons involving codes and economic development. A bench traces the route and shape of the Thames from Blackfriars to Dagenham. On its back are facts presenting the reality of the Thames, as it has increasingly become a site for various projects, with due economic and demographic impacts. Most local residents are forced away from their homes along the banks.

A 3D soundscape by world acclaimed sound artist and musician Martyn Ware immerses the visitor's imagination in the sounds, stories and sensations of the Thames. In between these two poles of the imaginary and the real, a series of commissioned films suggest an intermediary space of everyday life that still draws on the power of the Thames for inspiration.

The intention in presenting these multiple views of the Thames is to remind visitors that rivers everywhere have to be preserved in all their mysterious, mundane and mythic guises, and not be allowed to become commodified and exploited for their picturesque qualities.

P5

展项名称：当代中国的多元建筑实践
Exhibition Title: Diverse Practices in Contemporary Chinese Architecture

策展人：李翔宁，高长军
Curators: LI Xiangning, GAO Changjun

参展人：王澍等 33 位建筑师
Exhibitors : WANG Shu and other 32 architects

建筑师在当下面临着愈加丰富的时代背景以及愈显严峻的现实考验，挑战与机会并存。在历史、乡村、居住、新技术等诸多议题前，当代中国建筑师以多种姿态介入，从不同维度发声，令中国当代建筑呈现多元状态。从"权宜之计"到实用主义，从实验建筑到本土营造，从洋为中用到回归传统，当代中国建筑业已抵达总结前数十年实践的临界点。本展位通过回顾来展望下一个十年的建筑机遇，选取近年来最具代表性的更新改造类、乡村建设类、居住住宅类、数字化设计类等共三十余件建筑实例，从多维的视角展现当代中国建筑的丰富图景，提供对行业及社会总体形势的鸟瞰。

Contemporary architects are confronted with a more diverse background of the era and tougher reality. Challenges and opportunities currently coexist. Facing such issues as history, rural problems, habitation, new technology, contemporary Chinese architects get involved with practice in multiple approaches. From "makeshift" to pragmatism, from experimental architecture to fabrica loci, from cultural assimilation to Retour à tradition, contemporary Chinese architecture has arrived at the threshold to review the practice in the past decades. Looking further, this exhibition presents the most significant projects, covering architectural renovation, rural construction, housing and digital design. The more than 30 exhibited items form a parade of contemporary Chinese architecture and a bird's eye view of the profession and the society at large.

居住 01　李虎 + 黄文菁 Open Architecture 退台方院 B01 Stepped Courtyards / LI Hu + HUANG Wenjing, Open Architecture

居住 02　刘家琨 家琨建筑设计事务所 西村·贝森大院
B02 West Village·Basis Yard / LIU Jiakun, Jiakun Architects

居住 03　刘晓都 + 孟岩 都市实践 土楼公舍
B03 Tulou Collective Housing / LIU Xiaodu, MENG Yan, URBANUS

居住 04 马清运 马达思班 父亲住宅
B04 Father's House / MA Qingyun, MADA s.p.a.m

居住 05 陶磊 TAOA 凹舍　B05 The Concave House / TAO Lei, TAOA

居住 06 王灏 润·建筑工作室 砖宅农舍
B06 Wang House / Wang Hao, Rn Atelier

居住 07 谢英俊 谢英俊建筑师事务所 + 第三建筑工作室 杨柳村灾后重建
B07 Post-disaster reconstruction of Yangliu Village / Hsieh Yingchun,
Hsieh Ying-Chun Architects + Third Architectural Office

改造 01 葛明 东南大学 微园
C01 Wei Yuan Garden / GE Ming, School of Architecture,
Southeast University

改造 02 韩文强 建筑营设计工作室 胡同茶舍—曲廊院
C02 Tea House in Hutong / HAN Wenqiang, ARCHSTUDIO

改造 03 何哲, 沈海恩, 臧峰 众建筑 内盒院 C03 Courtyard House Plugin / HE Zhe, James Shen, ZANG Feng, Peoples' Architecture Office Co., Ltd

改造 04 刘珩 南沙原创建筑设计工作室 2013 深圳双年展浮法玻璃厂改造主入口 C04 Floating Entrance, Fufa Glass Factory Renovation / Doreen Heng LIU, NODE Architecture & Urbanism

改造 05 郭锡恩 + 胡如珊 如恩设计研究室 水舍 C05 The Waterhouse at South Bund / LYNDON NERI + ROSSANA HU, Neri&Hu Design and Research Office

改造 06 水雁飞 直造工作室 上海南外滩仓库办公改造 C06 1178 Waima Road Warehouse Renovation / SHUI Yanfei , MA Yuanrong , SU Yi-Chi, Naturalbuild

改造 07 王辉 都市实践 山西运城五龙庙环境整治设计 C07 The Five-Dragons Temple Conservative Landscape Design / WANG Hui, URBANUS

改造 08 王硕 META- 工作室 水塔展廊 C08 Water Tower Pavilion / WANG Shuo, META-Project

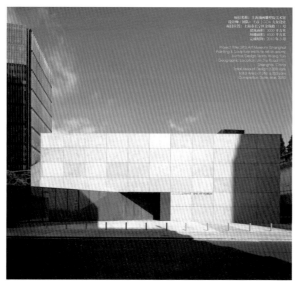
改造 09 王彦 GOA 大象设计 上海油画雕塑院美术馆 C09 Shanghai Painting & Sculpture Institute Art Museum / WANG Yan, GOA Architects

改造 10 曾群 巴士一汽四平路停车楼改造 C10 TJAD new office building / ZENG Qun, Tongji Architectural Design and Research Institute

改造 11 张轲 标准营造 微胡同 C11 Micro Hutong / ZHANG Ke, ZAO/standardarchitecture

改造 12 庄慎 阿科米星建筑设计事务所 衡山坊 8 号 C12 Façade renovation for No.8 building at Lane 890 Hengshan road / ZHUANG Shen, Atelier Archmixing

乡村 01 陈浩如 山上建筑工作室 太阳公社竹构系列·临安太阳公社 D01 Taiyang Organic Farming Commune / CHEN Haoru, Citiarc

乡村 02 何崴 三文建筑 / 何崴工作室 平田村爷爷家青年旅社 D02 Papa's Hostel / HE Wei, 3andwich Design / HE Wei Studio

乡村03 孔锐+范蓓蕾 亘建筑 齐云山树袋屋
D03 Treehouse at Qiyunshan / KONG Rui, FAN Beilei, genarchitects

乡村04 李晓东工作室 篱苑书屋 D04 Wattle School /
LI Xiaodong, LI Xiaodong Atelier

乡村05 刘宇扬 刘宇扬建筑事务所 景会设计 兴坪云庐度假酒店
D05 XY Yunlu Resort Restaurant Building / LIU Yuyang, Atelier LIU Yuyang Architects, Ares Partners

乡村06 王澍 业余建筑 杭州洞桥镇文村美丽宜居示范村
D06 Wencun Village / WANG Shu, Amateur architecture studio

乡村07 徐甜甜 DnA 多维度事务所 松阳红糖工坊 D07 Brown sugar workshop / XU Tiantian, DnA_Design and Architecture

乡村08 张雷 张雷联合建筑事务所 云夕深澳里书局一期工程
D08 Ruralation Shenaoli Library / ZHANG Lei, AZL Architects

乡村 09 赵扬 赵扬建筑工作室 标准营造 尼洋河游客中心
D09 Niyang River Visitor Center / ZHAO Yang, ZHAOyang Architects+ZAO/standardarchitecture

数字 01 马岩松，党群，早野洋介 MAD 哈尔滨大剧院 E01 Harbin Opera House / MA Yansong, DANG Qun, Yosuke Hayano, MAD Architects

数字 02 邵韦平 北京市建筑设计研究院 凤凰国际传媒中心 E02 Phoenix Center / SHAO Weiping, Beijing Institute of Architectural Design (Group) Co.

数字 03 宋刚，钟冠球 竖梁社 佛山艺术村建筑及景观设计 E03 Foshan Art Village / SONG Gang, ZHONG Guanqiu, Atelier cnS

数字 04 王振飞 华汇设计 2014 青岛世界园艺博览会天水综合服务中心、地池综合服务中心 E04 Service Center of International Horticultural Exposition 2014 Qingdao/ WANG Zhenfei, HHD_FUN

数字 05 袁烽 创盟国际 Archi-union 上海西岸 Fab-UnionSpace
E05 Fab-Union Space / Philip F. Yuan, Archi-union Architects

与水共生：世界优秀滨水空间案例展
Living with Water: World Outstanding Waterfront Space Case Exhibition

与水共生：世界优秀滨水空间案例展探讨了滨水空间成为城市公共生活的发生地的可能性，通过诸多世界滨水空间优秀案例的展示，关注了后工业城市滨水地带的再生与激活和气候变化下的弹性水岸的设计。

城市建立于江河湖海边，已经成为了一个常识，城市与水的关系梳理所当然的成为了每个城市无法回避的话题。在上海两岸贯通工程即将完成之际，滨水空间案例展选择这些世界级水岸作为展示案例，意在进行对比与展望。这些案例包括了威尼斯建筑大学、居依·诺丁森教授、BIG对全球气候变暖导致的海平面上升而提出的应对性策略和创造性设计，也有JCFO、捷得、SASAKI、维斯/曼弗雷迪建筑·城市·景观·设计事务所，West 8对城市滨水空间的重新激活，还有BAUM建筑师对城市现存水岸的批判性阅读。在此，我们将它们进行集中展示，同时希望人们获得属于自己的对于城市滨水空间的期待。

策展人：李翔宁，莫万莉，邓圆也，张子岳
Curator: LI Xiangning, MO Wanli, DENG Yuanye, ZHANG Ziyue

Living with Water: World Outstanding Waterfront Space Case Exhibition discusses multiple possibilities of making waterfront space a place for the happening of public lives. By demonstrating several world outstanding waterfront cases, the exhibition focuses on the revitalization of post-industrial waterfront and the design of resilient waterfront in respond to climate change.

It is common knowledge that cities are usually built by rivers or lakes. Thus, the relationship between urban space and waterfront space becomes a critical matter to cities around the world. As Shanghai's Continuous Waterfront project is about to finish, the exhibition chooses several world-class waterfront space as cases for comparison and outlook. These cases include responsive strategy and creative design proposals by IUAV, Guy Nordenson and Associates and Bjarke Ingels Group, as well as design projects by James Corner Field Operations, Jerde Partnership, SASAKI, WEISS/MANFREDI Architecture/Landscape/Urbanism, West 8 and a critical reading of existing urban waterfront space developed by BAUM Architects. By putting them together, the exhibition hopes to offer visitors a variety of visions for future urban waterfront space.

展项名称：与水共生：世界优秀水岸空间案例展
Exhibition Title: Living with Water: World Outstanding Waterfront Space Case Exhibition

策展人：李翔宁，莫万莉，张子岳，邓圆也
Curators: LI Xiangning, MO Wanli, ZHANG Ziyue, DENG Yuanye

参展人：BAUM建筑师事务所，B.I.G，居依·诺丁森结构设计有限公司，JCFO，SASAKI，威尼斯建筑大学，维斯，曼弗雷迪建筑·城市·景观·设计事务所，West 8
Exhibitors : BAUM Architects, Bjarke Ingels Group, Guy Nordenson and Associates, James Corner Field Operations, SASAKI, Universita IUAV di Venizia, WEISS , MANFREDI Architecture · Landscape · Urbanism, West 8

《与水共生：世界优秀滨水空间案例展》探讨了滨水空间成为城市公共生活的发生地的可能性，通过诸多世界滨水空间优秀案例的展示，关注了后工业城市滨水地带的再生与激活，以及气候变化下的弹性水岸的设计。在上海两岸贯通工程即将完成之际，滨水空间案例展选择这些世界级水岸作为展示案例，意在进行对比与展望。这些案例包括威尼斯建筑大学、居依·诺丁森教授、BIG对全球气候变暖导致的海平面上升而提出的应对性策略和创造性设计，也有JCFO、SASAKI、维斯／曼弗雷迪建筑·城市·景观·设计事务所、West 8对城市滨水空间的重新激活，还有BAUM建筑师对城市现存水岸的批判性阅读。在此，我们将它们进行集中展示，同时希望人们获得属于自己的对于城市滨水空间的期待。

Living with Water: World Outstanding Waterfront Space Case Exhibition discusses multiple possibilities of making waterfront space a place for the happening of public life. Showcasing extraordinary waterfront cases around the world, the exhibition focuses on revitalizing a post-industrial waterfront and the design of a resilient waterfront in response to climate change. As Shanghai's Waterfront Connection is about to finish, we choose several world-class waterfront space to evaluate how Shanghai has performed and what it can do from now on. These cases include a responsive strategy and creative design proposals by IUAV, Universita IUAV di Venizia and Bjarke Ingels group, as well as designs by James Corner Field Operations, SASAKI, WEISS/MANFREDI Architecture/Landscape/Urbanism, West 8 and a critical reading of existing urban waterfront spaces developed by BAUM Architects. By putting them together, we hope to offer visitors a future vision of urban waterfront spaces.

展项名称：MOSE 计划 / 参展人：威尼斯建筑大学
Exhibition Title: MOSE Project / Exhibitors: Università IUAV di Venezia

展项名称：涟漪效应 / 参展人：SASAKI
Exhibition Title: Ripple Effect / Exhibitors: SASAKI

展项名称：芝加哥海军港 / 参展人：JCFO
Exhibition Title: Chicago Navy Pie / Exhibitors: James Corner / Filed Operation

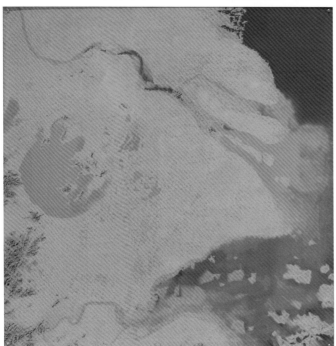

展项名称：气候变化与弹性滨水空间 / 参展人：居依·诺丁森结构设计事务所
Exhibition Title: Climate Change Waterfront Resiliency / Exhibitors: Guy Nordenson & Associates

展项名称：曼哈顿大 U / 参展人：B.I.G
Exhibition Title: Manhattan Big U / Exhibitors: Bjarke Ingels Group

展项名称:马德里曼塞纳雷斯河岸更新景观工程 / 参展人:West 8
Exhibition Title: Madrid Rio: A Project of Urban Reconnection / Exhibitors: West 8

展项名称：斯科普里的整容手术 / 参展人：BAUM architects
Exhibition Title: Skopje's Plastic Surgery / Exhibitors: BAUM architects

展项名称：西雅图奥林匹克雕塑公园与康奈尔大学纽约校区 / 参展人：维斯 / 曼弗雷迪建筑·城市·景观·设计事务所
Exhibition Title: Seattle Olympic Park and Cornell NYC / Exhibitors: WEISS / MANFREDI Architecture / Landscape / Urbanism

展项名称：社会图景：来自城市内部的影像学
Exhibition Title: Social View: Iconography from City

策展人：金江波
Curator: JIN Jiangbo

参展人： 敖国兴，戴建勇，冯梦波，何崇岳，金江波，吉姆·斯皮尔斯、李消非，李振宇，马良，倪卫华，Nancy Royal，渠岩，沈少民，邵文欢，王川，王庆松，徐坦，杨泳梁，赵宏利，曾力
Exhibitors: AO GuoXing, DAI Jianyong, FENG Mengbo, HE Chongyue, JIN Jiangbo, Jim Speers, LI Xiaofei, LI Zhenyu, MA Liang, NI Weihua, Nancy Royal, QU Yan, SHEN Shaomin, SHAO Wenhuan, WANG Chuan, WANG Qingsong, XU Tan, YANG Yongliang, ZHAO Hongli, ZENG Li

P7

本展览作为2017上海城市空间艺术季的展示板块之一，参展的大部分艺术家均活跃于国内外各项重要展览活动，他们以丰富的视觉经验与文化游历，试图通过犀利的镜头，以其先锋的艺术姿态，讨论当代摄影与现实城市的关系。

此次展览的作品以呈现中国当代社会进程中所独有的视觉范本为例，以艺术家的立场构筑观察的维度，探讨影像学定义下城镇化中的社会景观。参展艺术家与作品以其独特的艺术性超越了影像的新闻性和纪实性，在揭开影像的表象之时，希冀观者关注到时间、人、空间和现场的诸多关系，发掘人文环境里被忽略的内涵特质，提示观者们以主动的方式审视城市的不同方位，重新思考城市文明前行的方向。

As an exhibition section of SUSAS 2017, it features works of artists actively engaged in major exhibitions around the world. With rich experiences, visually and culturally, gained from their tours, they try to discuss the relation between contemporary photography and real cities through keen camera lens in an Avant-garde pose. Here, you will see Unique visual recordings of social processes going on in contemporary China wherein artists develop their own perspectives and discuss social spectacles during urbanization ichnographically.

The artists and their works have transcended the journalism and documentary nature of images with unique artistry. Unveiling images, they expect viewers to focus on various relations between time, people, space and scene, and exploits the neglected connotations and essence of built environments. Meanwhile, they remind viewers to inspect different cities' orientations and rethink where urban civilization is heading for.

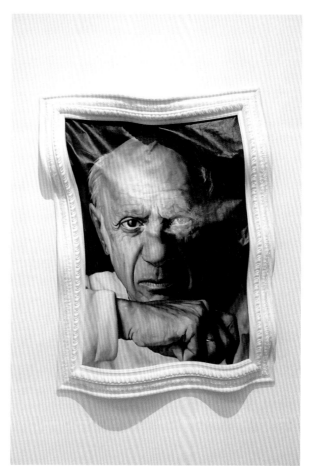

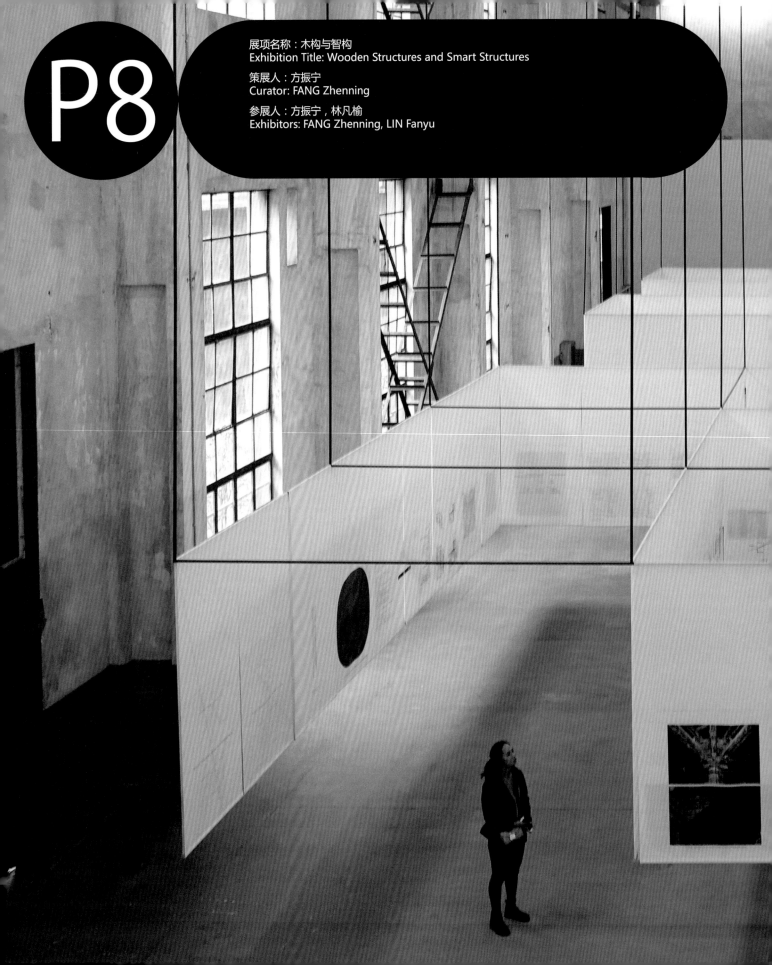

P8

展项名称：木构与智构
Exhibition Title: Wooden Structures and Smart Structures

策展人：方振宁
Curator: FANG Zhenning

参展人：方振宁，林凡榆
Exhibitors: FANG Zhenning, LIN Fanyu

"木构"就是木结构，"智构"就是智能结构。把"木构"与"智构"并列起来，是探讨传统建筑的木结构和现代智能结构之间的继承关系。我们发现中国建筑在几千年前，就创造了模件体系。木构系统的精粹实际上是预制，木构中的木构件，如同中国象形文字中的偏旁部首一样，它是靠架构搭建一幢房子。

这一研究是我受第十四届威尼斯国际建筑双年展总策展人雷姆·库哈斯的邀请，参与"建筑元素"部分的OMA工作坊，解读《营造法式》提交的研究成果。该成果的核心部分曾在双年展现场发表，而现在的展示只是对发表部分的延伸阅读。我们为什么要研究一本九百年前中国建筑的法规《营造法式》？不只是因为它是一本在时代和建筑方面都位于巅峰期的包罗万象的建筑经典，更重要的是它让我们有机会从历史的智慧中发现孕育现代性的可能。如果《营造法式》是一本了解中国建筑的文法书的话，那么模件体系就是这本书中"有机"的部分。实际上中国木构建筑的建造过程是一个生产的过程，正是由于它的预制系统，才能有可观的建造速度和大规模的普及。中国木构建筑的梁柱体系属于有机建筑，而来自纽约的激浪派智能的预制生命系统，可以说是《营造法式》中模件体系的现代版。

By placing "wooden structures" and "smart structures" in parallel discourse, we investigate the relationship between traditional wood architecture and modern smart systems. We discovered that modular systems were already invented in Chinese architecture thousands of years ago. The essence of wooden structures lies in prefabrication: each component in a wooden structure is like an element in an ancient Chinese pictogram, it is the sum of these parts that creates a framework for architecture.

I was invited by Mr. Rem Koolhaas, the Chief Curator of the 14th Venice Biennale International Architecture Exhibition to participate in the part of "Architectural Elements" in the OMA Workshop: research conducted for *Yingzao Fashi*. The main part of the report has been presented at the Biennale and this time I will have an extended interpretation based on it. Why do we study a Chinese architectural treatise published 900 years ago? Not only is it an architectural classic produced at the height of an era flourishing with architectural advancements, but more importantly, it is a historical guidebook with potential knowledge for modern application. Technically, the construction process of Chinese wooden structures is manufacturing in the modern sense since it is created as a prefabricated system with superior construction efficiency and aptness for mass production. If we consider the framed construction of beams and columns in wooden Chinese structures as organic architecture, then the Fluxus smart-prefab living system from New York can be understood as a modern version of the modular systems described in *Yingzao Fashi*.

《筑建图》/ 南宋 / 佚名 / 龙美术馆 藏
ZHUJIANTU / The Southern Song Dynasty / anonymous / Long Museum Collection

SECTION PERSPECTIVE OF WOOD STRUCTURES
木结构剖透视图

SECTION PERSPECTIVE OF FLUXUS PREFAB SYSTEM
激浪派预制体系剖透视图

VENTILATION SYSTEM
通风系统

HOLLOW STRUCTURAL COMPONENTS
空心结构部件

PIPING SYSTEM
管道系统

SEALING SYSTEM
密封系统

LIGHTING SYSTEM
照明系统

SEALING SYSTEM
密封系统

ELECTRICAL SYSTEM
电路系统

MULTIFUNCTIONAL STRUCTURAL COMPONENTS
多功能结构部件

ELECTRICAL SYSTEM
电路系统

HOLLOW STRUCTURAL COMPONENTS
空心结构部件

PIPING SYSTEM
管道系统

VENTILATION SYSTEM
通风系统

P9

展项名称：漫步环翠堂园景
Exhibition Title: Stroll in Huancui Tang

策展人：方振宁
Curator: FANG Zhenning

参展人：方振宁
Exhibitor : FANG Zhenning

《环翠堂园景图》是明代万历年间徽派版画的代表作，也是中国长卷绘画作品中独特的长卷作品，是继宋《清明上河图》之后又一幅珍贵的、具有很高价值的艺术作品。作者以浪漫、自由、理想化的手法，表现了令人神往的明代园林艺术。这幅作品虽然是一幅木版画，却是少有的木版画长卷，画中无论是风景还是人物以及建造物，都刻画得非常精致和生动，让我们看到了明朝中后期江南六府经济繁荣，古典园林艺术水平达到的顶峰状态。我们将原本只有24厘米高、14米长的版画，放大到120厘米高、108米长的画面，让观众感觉自己就在漫步园林。

HuanCui Tang landscape is one of the most famous engravings from the Ming dynasty in the Hui Style, making it another treasure that ancient Chinese art has to offer after *Along the river during the Qingming Festival*. By using romantic, free and idealized visual language, its author demonstrates a fascinating classical garden of the Ming dynasty. The landscapes, figures and buildings are drawn delicately and lively throughout the work whereby we should have a glimpse of the prosperity of south China and classical Chinese garden-building at its height. For the sake of exhibition, we have enlarged the scale of the printmaking from 24x1400 cm to 120x10800 cm to create an environment that makes people feel like they're really strolling in Huancui Tang.

116

P10

展项名称：马列维奇视觉年表
Exhibition Title: Malevich Visual Chronology

策展人：方振宁
Curator: FANG Zhenning

参展人：方振宁
Exhibitor: FANG Zhenning

作为 20 世纪最重要的艺术家之一和至上主义的创始人，卡济米尔·马列维奇（1879—1935）对 20 世纪的全球艺术产生了不可估量的影响，这种影响是在我们对现代艺术流派的深入研究之后逐渐显示出来的。与马列维奇同时代的先锋艺术家不乏其人，但为什么只有马列维奇一人有那么大的影响力？这是因为他的作品独具革新性和精神性，所以才成为不只是俄罗斯，也是世界艺术宝库中的瑰宝。

编者对马列维奇的资料收集和研究长达三十年，在这个基础上编辑的马列维奇视觉年表，是大数据方法论的成果，也是中文圈首次系统总结和梳理马列维奇的至上主义遗产。这一年表是在 2014 年马列维奇的至上主义宣言发表一百周年时首次编辑的年表的基础上的增订版。

Kazimir Malevich (1879—1935), an important artist and pioneer of Suprematism, who had a great influence on the art community in the 20th century, and we only gradually grasp his significance after in-depth studies of contemporary art. Why did Malevich have such influence among and beyond so many avant-garde artists in his time? The reason why his work is treasured both in Russia and worldwide is due to its innovation and spiritual nature.

The editor has collected and researched information on Malevich for around thirty years. Malevich Visual Chronology is based on a thorough survey. It is also the first systematic summary and analysis of Malevich's suprematist works in the Chinese academic community. This chronology is a revised version based on the first version which was composed for the 100th anniversary of the Malevich Suprematism Statement in 2014.

彼得格勒"最后的未来主义画展 0,10"中的马列维奇作品 /1915

黑色至上主义方块 /79.6x79.5cm/ 布面油画 /1915

莫斯科特列恰科夫画廊马列维奇作品陈列

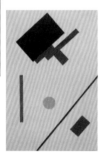

至上主义构图 /27.5x19cm/ 布面油画 /1915

无题 /11x11.4cm/ 铅笔纸本 /1915

至上主义构成 /21.5x14cm/ 铅笔纸本 /1915

无题 /17x12.5cm/ 铅笔纸本 /1915

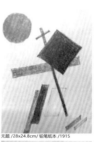

无题 /28x24.8cm/ 铅笔纸本 /1915

至上主义 /97x66cm/ 布面油画 /1915

构成 11r/11.2x16cm/ 铅笔纸本 /1915

飞机的至上主义绘画 /58.1w48.3cm/ 布面油画 /1915

至上主义 /53.3x53.7cm/ 布面油画 /1915

至上主义：抽象 /80x80cm/ 布面油画 /1915

构成 /16.3x11.1cm/ 铅笔纸本 /1915

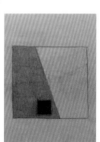

至上主义：倾斜平面上的正方形 /15.4x11cm/ 铅笔墨汁纸本 /1915

建造中的房子绘画 /15.4x11.2cm/ 铅笔纸本 /1915

至上主义的结构 /11.2x16.3cm/ 铅笔纸本 /1915

无题 /50.5x43cm/ 铅笔纸本 /1915

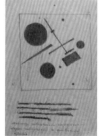

圆圈和正方形 /15.8x10.8cm/ 铅笔纸本 /1915

有斜边物体的长方形 /12.5x11.6cm/ 铅笔纸本 /1915

1915

在展厅的"红角",即斯拉夫人通常悬挂圣像画的位置,马列维奇放置上《黑色正方形》,这幅画作随后成为俄国先锋派最著名的画作,被后人称为"俄国先锋派圣像"。展览上还展出了马列维奇的其他画作,包括《黑色十字》《黑色的圆》《红色方块》以及他同行和学生的画作,共计39幅。展览所得的一半利润被归入了艺术工作者医院。

"0,10"展在当地艺术圈引起了极大的反响,虽然在今日,大多数艺术史学者将此展看作是俄国先锋派的伟大成就,但在当时它主要引来的是质疑之声。艺术家亚历山大·别努阿就言辞犀利地批评了该展,尤其是《黑色正方形》这幅作品,很多批评家对于马列维奇将这幅画放置在通常悬挂圣像画的角落一事极其不满。

1915年12月2日至16日《征服太阳》的设计展在俄罗斯纪念剧场展出。

红色方块 /53x53cm/ 布面油画 /1915

红色正方形 /6.1x5.7cm/ 水粉纸本 /1915

结构 31/75x75cm/ 布面油画 /1915

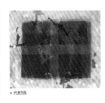
x 光射线

黑色四边形 /17x24cm/ 布面油画 /1915

黑白至上主义构成 /80x80cm/ 布面油画 /1915

短长的飞机 /80x80cm/ 布面油画 /1915

黑色正方形 /49x49cm/ 布面油画 /1915

至上主义构图 /80.4x80.6cm/ 布面油画 /1915

至上主义 /53x53cm/ 布面油画 /1915

至上主义运动中的绘画体积 /87.5x72cm/ 布面油画 /1915

至上主义 /101.5x62cm/ 布面油画 /1915

三个不规则四边形 /17.3x20.7cm/ 铅笔纸本 /1915

马列维奇展1973年11月16日—1974年1月13日

1915

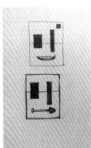
构成 /16.5x11.1cm/ 铅笔纸本 /1915

无题 /10.2x7.7cm/ 铅笔纸本 /1915

至上主义构成 /1915

至上主义 /70x60cm/ 布面油画 /1915

围绕着长方形和三角形的环 /11.2x16.6cm/ 铅笔纸本 /1915

构成 /11.2x16.5cm/ 铅笔纸本 /1915

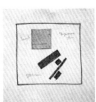
构成 21c /9.7x9.7cm/ 铅笔纸本 /1915

黑色正方形和白色长方形以及红色笔触 /12.5x9.7cm/ 铅笔纸本 /1915

至上主义构成 /铅笔水彩纸本 /1915

构成 /11.2x16.5cm/ 铅笔纸本 /1915

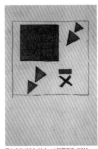
至上主义 /16.5x11.1cm/ 铅笔纸本 /1915

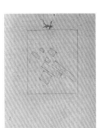
八个红色长方形 /1915

四边形和圆 /44.4x31.2cm/ 布面油画 /1915

四边形和圆 分析图

四个立方体和一个黑色梯形 /11.9x11.1cm/ 铅笔纸本 /1915

深蓝色三角形与黑色矩形 /57x66.5cm/ 布面油画 /1915

至上主义构图:飞机投影 /57x57cm/ 布面油画 /1915

黑色和白色的管状边缘 / 铅笔纸本 /1915

P11

展项名称：万象
Exhibition Title: Social View: Wanxiang

策展人：方振宁
Curator: FANG Zhenning

参展人：赵弘君，何汶玦，靳烈，王欣，马岩松，王昀，孟禄丁，李迪，李磊，童振刚，方振宁，陈文令，吕越，张朝晖，FANGmedia
Exhibitors: ZHAO Hongjun, HE Wenjue, JIN Lie, WANG Xin, MA Yansong, WANG Yun, MENG Luding, LI Di, LI Lei, TONG Zhengang, FANG Zhenning, CHEN Wenling, Lyu Yue, ZHANG Zhaohui, FANGmedia

这个以"万象"为主题的策划，为十几名艺术家、建筑师、出版人和策展人提供一个和观众互动的机会。我们的互动方式非常简单，就是给每位参加者一个尺寸定为 635mm x 965mm 即一张全开纸大小的展出发表机会，也就是把大家的作品印刷在这张全开纸张上，印刷数量为一万张，这一万张印刷品整齐地排列在会场中，参观者可以根据自己的爱好随便带走。我们把它看作是一种特别的传播方式，也可以视为纸上美术馆。

In the name of Wanxiang, meaning variety, this exhibition provides 20 artists, architects and publishers an opportunity to interact with the viewers. The interaction is quite simple: each artist is given an opportunity to publish their works, which will be printed on 10,000 sheets of paper (size of 635mm x 965mm). Then these paper sheets will be presented in order at the exhibition so that visitors can choose their favourite. This is valued as a special way of communication and can also be interpreted as a gallery on paper.

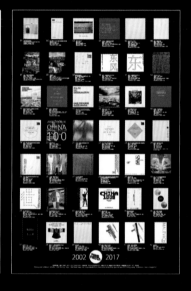
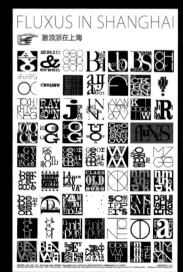

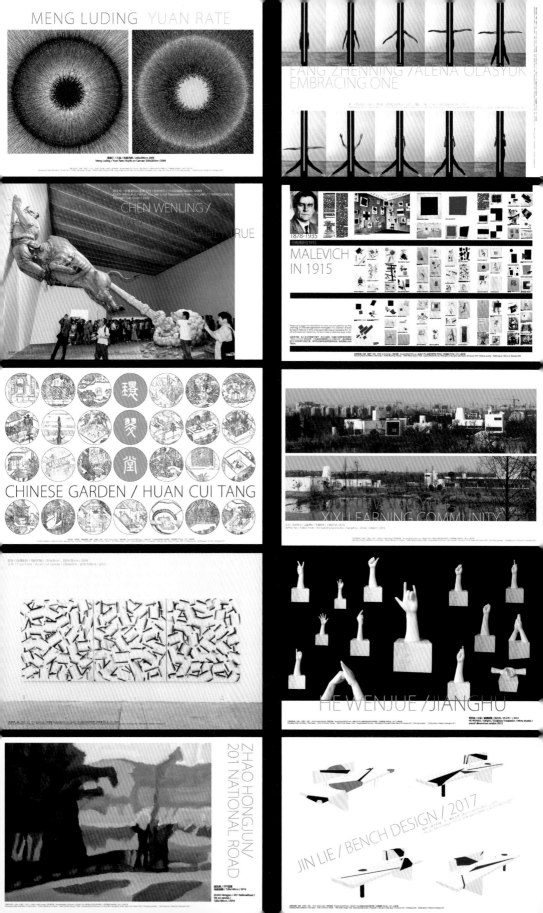

鼓浪屿位于福建省厦门市西南,是一座面积仅1.88平方千米的海岛。从19世纪中叶至20世纪中叶,鼓浪屿通过闽南本土居民、外来多国侨民和还乡华侨群体的共同营建,经历了从传统聚落到殖民风格居留地,再到兼具国际化与本土化特征的现代社区的跨越发展,成为具有突出文化多样性和近代化生活品质的国际社区。一个小岛,通过与世界的连接、包容和吸纳,创造出自己独有的文化景象。

今天的鼓浪屿成为国内最受欢迎的旅游目的地之一,但也同样面临着旅游发展的压力和社区成员更替、活力衰退等挑战。从共享遗产的视角重新审视鼓浪屿的独特价值,理解不同文化群体在历史国际社区发展中的贡献,以及由本地和外来、传统和现代文化的交融所形成的鼓浪屿文化特质,使文化遗产保护逐渐成为政府和社区民众的共识,以此作为鼓浪屿应对挑战、恢复社区文化自信、激发社区活力的重要途径。

2017年7月8日,"鼓浪屿历史国际社区"正式列入世界文化遗产名录,参展团队作为鼓浪屿申遗全程技术支持方,以"共享遗产保护之路"为主题与大家分享鼓浪屿遗产保护的经验。展览以鼓浪屿遗产保护为视角,围绕历史建筑、城市公共空间、社区生活三个主题展开,通过历史照片与现状照片的比对,讲述以社区和人为核心的遗产保护工作如何帮助鼓浪屿重续断裂的文化,建立物质遗产与社会生活的连接,政府与社区民众的连接,以及鼓浪屿历史、当代与未来的连接。

Located off the southwest coast of Xiamen, Fujian Province, Kulangsu is a small island with an area of merely 1.88 square kilometres. From the mid-nineteenth to mid-twentieth century, with the joint efforts of local residents, returned overseas Chinese, and multi-national immigrants, Kulangsu experienced a leaping development from a village to a small colony, and then to a major international settlement with diversified cultures and modern living standards. Through connections and communications with the outside world, Kulangsu witnessed interaction, collision and fusion of diversified cultures, which helped the island establish its unique cultural identity.

Today, Kulangsu has become one of China's most popular tourist destinations. As with the same challenge that most tourist spots are facing, Kulangsu is under the pressure of roaring tourism and declining community vitality. Re-examining Kulangsu's unique value from the perspective of Shared Built heritage, it is critical for both the government and the public to understand the contributions different cultural groups have made in developing the historic International Settlement, as well as the establishment of Kulangsu's cultural characteristics integrated by local and foreign, traditional and modern cultures. This could be regarded as a practical way for the government and community to reach consensus on cultural heritage conservation. This should be an important approach to restore confidence and inspire vitality in the community.

On July 8th, 2017, "Kulangsu, a historic International Settlement" was officially a World Cultural heritage. As the technical support team of Kulangsu's inscription nomination process, we are delighted to share Kulangsu's heritage protection experiences given the theme "Conserving Shared Built heritages". Through the perspective of heritage protection, the exhibition focuses on three topics: historic buildings, urban public spaces, and community life. Each topic is demonstrated through contrasting photographs of past and present, as a narration of how community and resident-centered heritage protection has helped restore Kulangsu's fragmented culture and establish connections between tangible sites and social life, government and community, and Kulangsu's past, present and future.

公共空间形态
TOPOLOGY

城市公共空间历来具有丰富多样的形态。在当代，传统类型的公共空间依然具有重要的作用，而与此同时，公共空间也由于新媒体、网络数字技术和生活方式的改变而呈现出新的面貌，给现有的城市和建筑环境提出了新的挑战。新型和传统形态的公共空间的混合，将是未来发展的方向。我们从物体与媒介两个层面上选择出各项展品，观察并比较它们之间的差异和转换，通过建筑、雕塑、装置和话语，来探讨和表达混合多样的公共空间形态。

Urban public space has a variety of forms. In the contemporary era, the traditional type of public space still plays an important role. At the same time, public space is also showing a new look due to changes in new media, digital technologies and Internet lifestyle, which presents the existing urban and architectural environment with new challenges. Mixture is the trend of public spaces. The structures, sculptures, installments and discourse projects exhibited here, whose divergences and transitions are carefully observed and compared, are chosen at the levels of both objects and media whereby we explore and express mixed and varied forms of public space.

T1

展项名称：林中之境
Exhibition Title: Radura

策展人：斯坦法诺·博埃里
Curator: Stefano Boeri

参展人：斯坦法诺·博埃里
Exhibitor: Stefano Boeri

在森林中，空地具有非常重要的作用，它们为生物多样性的扩散和繁殖提供了条件。

在城市环境之中，Radura 有着同样的意义：它是一种公共空间的原型，为大都市的快速流动提供纾解。它让我们有机会在忙碌的都市生活中慢下来，同时创造了一个混合空间，让男女老少、动物人类在此与自身、与彼此重新建立起联系。

正是由于它的混合特征，Radura 也成为一次机遇，能够在自然与人类环境之间创造出强有力的空间及象征性联系。和在森林里促进生态多样性的空地一样，城市环境中的 Radura 也能够营造出一个"悬挂空间"，让人们在此停留、休憩、等待。我们需要退后一步，从地质学的视角来观察地球，这样才能更好地理解 Radura 这样的空间的意义。

Inside forests, clearing plays a key role. As voids carved into a mass, they represent conditions for the proliferation and multiplication of biodiversity.

Inside the urban environment RADURA — the installation made up of a circle of 350 cylindrical wooden columns, by Stefano Boeri Architetti — has the same meaning: a prototype of a public space for decongestion within metropolitan flows. It begs an opportunity to slow down from the hectic life the metropolis pushes us to live while creating a hybrid space where man and women, children and animals can reconnect with themselves and with others.

Thanks to its hybrid nature, RADUTA represents an opportunity to create a strong spatial and symbolic connection between the natural and anthropic environments. As clearings in the forests foster biodiversity, RADURA in the urban context creates a suspended space in which to stop, rest, and wait within a generative and regenerative space. To better understand the meaning of a space such as RADURA, we need to take a step back and observe our planet from a geological perspective.

太空天线 2117
RADURA 2117

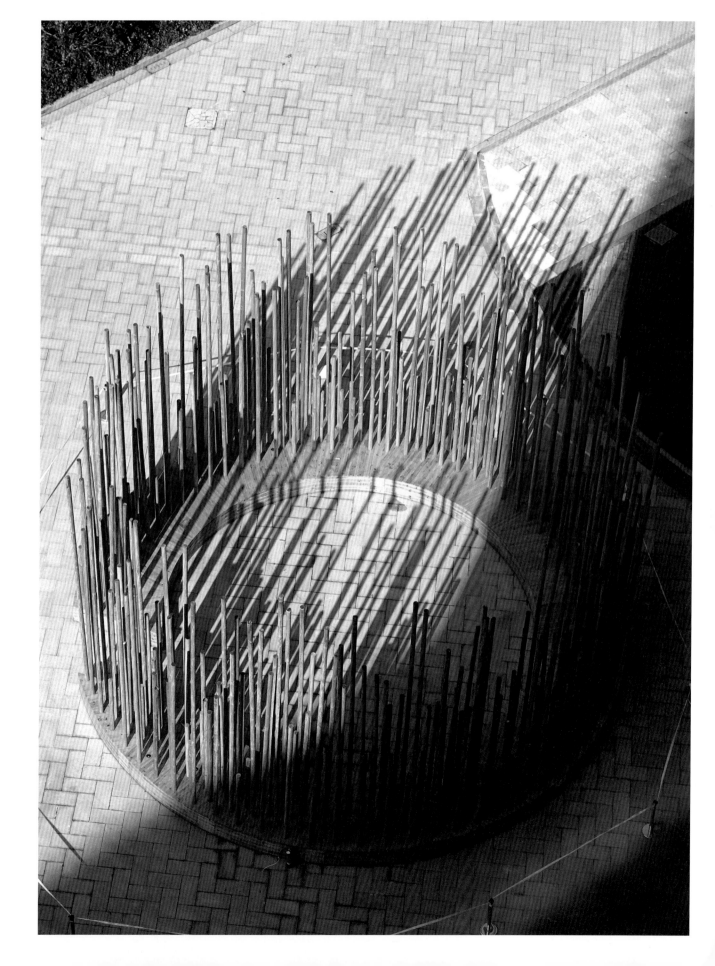

T2

展项名称：2017 年欧盟当代建筑奖——密斯·凡·德·罗奖和 2016 年 "Fear of Columns" 竞赛展览
Exhibition Title: 2017 The EU Mies Award and 2016 "Fear of Columns"

策展人：李翔宁，高长军
Curators: LI Xiangning, GAO Changjun

参展人：密斯·凡·德·罗基金会
Exhibitor: Mies van der Rohe Foundation

本次展览展示的是2017年欧盟当代建筑奖——密斯·凡·德·罗奖的部分成果，以及为德国馆重建30周年纪念所推出的"Fear of Columns"的竞赛成果。

2017年欧盟当代建筑奖——密斯·凡·德·罗奖的部分展示了最终入围的5个作品，包括最后的得主，其对公共空间的关注颇为亮眼。"Fear of Columns"的部分展示了当下对于1929年在巴塞罗那德国馆前建造的8根爱奥尼克柱的重新理解，是一次时间维度的新旧连接。

This exhibition presents the results of the 2017 EU Mies Award and the intervention ´Fear of Columns´ carried out on the occasion of the 30th anniversary of the Pavilion's reconstruction.

The EU Mies Award 2017 part shows the five works as well as the finalists, including the winner, which is concerned about public space. "Fear of Columns " part shows a reinterpretation for the 8 Ionic columns in front of the Barcelona Pavilion. Thus a new dimension in connecting time is created.

展项名称：全球建筑实践罗盘：一种新兴建筑的分类学
Exhibition Title: Architecture's "Political Compass": A Taxonomy of Emerging Architecture

策展人：亚历杭德罗·扎拉-波罗，吉尔莫·费尔南德兹-阿巴斯卡尔
Curators: Alejandro Zaera-Polo, Guillermo Fernandez-Abascal

参展人：亚历杭德罗·扎拉-波罗，吉尔莫·费尔南德兹-阿巴斯卡尔
Exhibitors: Alejandro Zaera-Polo, Guillermo Fernandez-Abascal

这一研究项目勾勒出了21世纪经济危机之后作为一种回应的新兴建筑学实践的现状，通过分类学，新的实践方式被加以分类和研究。"全球建筑政治事件"则是一系列延续上述研究的公开研讨活动。通过被列入图表的建筑师的参与和研讨，这一活动希望能够挑战现有的分类而形成迭代的图表。研讨活动于2017年在全球不同城市和机构展开，包括AA、ETH等。

The research project outlines a 21st-century taxonomy of architecture, attempting to define and categorize various new forms of practice that havegrown in popularity in the years since,and as a political response to the economic crisis. "global Architectural Political Events" are a series of public debates that continue the investigationabout political re-engagement of the discipline. Challenging the established categories, their relationships and featured protagonists, the expected outcome of this series is a new iteration of the diagram. The discussions took place across different cities and institutions, including AA, ETh and others around the world throughout 2017 with the participation of several offices featured on the map.

T4

展项名称：哥伦布之谜系列雕塑展
Exhibition Title: The Mysteries of Columbus Cristobal

策展人：戈勃朗基金会
Curator: Cristobal Gabarron Foundation

参展人：戈勃朗基金会
Exhibitor: Cristobal Gabarron Foundation

"哥伦布之谜"是一组由 10 件原创雕塑组成的作品。哥伦布是世界历史上最具有传奇色彩的开拓者之一,这组雕塑作品旨在描绘他的形象、才华、激情和奇幻经历。作品最初的灵感来源于哥伦布的真实身份与性格这一众说纷纭的千古之谜。以"哥伦布之谜"为主题的系列雕塑撷取了几件与哥伦布相关的历史事件,试图通过他本人的记叙来诠释他的思绪、抱负和梦想。作品使我们开始重新审视这位伟大探险家的生平,并且发现,许多流传甚广的哥伦布故事都是后人杜撰的,如果按照他们的描述,哥伦布就成了一个具有三四种迥异性格的严重人格分裂者。

"The Mysteries of Columbus" presents the creation of a sculptured work composed by 10 original sculptures, allegoric to the figure, talent, passion and magic of one of the greatest explorers in history. These sculptures have been created in a major essential mystery that involves the personality of the discoverer.

T5

展项名称：凝聚
Exhibition Title: Cohesion

策展人：方振宁
Curator: FANG Zhenning

参展人：方振宁
Exhibitor: FANG Zhenning

作品利用 257 库在特展区以外剩余的空间，特别利用一些不可移动的与建筑室内相关的设施，也就是对观赏有障碍作用的东西，将它粉刷改装成艺术装置，统称为室内公共空间中的艺术装置。我们把表达空间的语言凝练为几何形，这是 20 世纪艺术史上最重要的革命，也就是立体主义这一分析的体系推崇的语言，"凝聚"这个概念是向至上主义表达敬意的空间至上主义的当代版，由不规则立方体、三角柱、十字形和正方形等四件作品组成一组。

With the residual space of the 257 Factory, the artist paint the immovable indoor facilities, which used to be obstacles into installations . Following the most important revolution in the art of 20th century and , we use geometries to express the space. The concept of "Cohesion " is a contemporary tribute paid to suprematism, which alludes to ideas of irregular composition of cubes, triangular prisms, crosses and squares.

T6

展项名称：混乱中迷失
Exhibition Title: LOST IN A SHUFFLE
策展人：方振宁
Curator: FANG Zhenning
参展人：劳伦斯·维纳
Exhibitor: Lawrence Weiner

"LOST IN A SHUFFLE"是英文中的一句俚语，无论是字面还是隐喻含义，它均可以用于形容人在某个情境下不堪重负，或者是人故意将自己置于会迷失的情形下。圆周运动既是一个平面上的动作，也是一个文字内容的图像化表达，用以赞美迷失的表象下循环往复的本质。在这种方式下，劳伦斯·维纳提出了文字与图像的关系。在维纳的叙述中，所有的表达都经过仔细推敲，并且以几乎跳脱出情境的方式展示出来，带动观者自行解读作品的意义。

这件作品最早于2015年在Mai 36美术馆的"WITHIN GRASP"展览中以不同的排版、颜色和语言设计展出。根据不同的情境，作品可以在视觉上发生变化并适应。在这次的展览中，作品虽然改变了它的形式、色彩和语言，但保留了其概念的本质。

The expression LOST IN A SHUFFLE is an English saying that can be used when one is either overwhelmed in a situation or if one intentionally exposes themselves to an experience or situation that may carry them away -- in the literal or metaphorical sense. The circular motion is a graphic gesture and imagery of textual content, complementing the circular movement of a shuffling act. In this way, Weiner suggests a relation between text and imagery. All the phrases in Weiner's statements are carefully chosen and seemingly presented out of context, as their meaning is perceived by the beholder.

The work LOST IN A SHUFFLE was first exhibited at Mai 36 Galerie in 2015 during the exhibition WITHIN GRASP in a different layout and color and language design. It is through circumstances that a work can visually change and adapt, in this case a change of form, color and language, all the while keeping its essential concept.

INSTRUCTIONS FOR
MAI 36 GALLERIE
SHANGHAI URBAN SPACE ART SEASON 2017
OCTOBER 2017

ROUGH MOCKUP

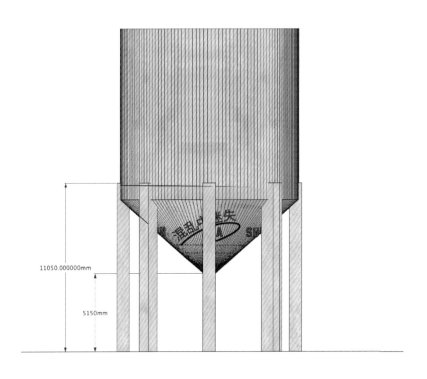

INSTRUCTIONS FOR
MAI 36 GALLERIE
SHANGHAI URBAN SPACE ART SEASON 2017
OCTOBER 2017

ROUGH DIMENSIONS

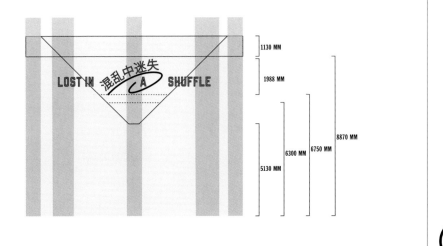

INSTRUCTIONS FOR
MAI 36 GALLERIE
SHANGHAI URBAN SPACE ART SEASON 2017
OCTOBER 2017

PLEASE USE THE FOLLOWING COLORS OR AS CLOSE AS POSSIBLE:
    RED: PANTONE 032 U
    BLACK: PANTONE PROCESS BLACK C

LOST IN 混乱中

LAWRENCE WEINER STUDIO
29 AUGUST 2017

迷失
A SHUFFLE

展项名称：风卷
Exhibition Title: Rolling Wind

策展人：方振宁
Curator: FANG Zhenning

参展人：王迈
Exhibitor: WANG Mai

T7

城市里的草坪和景观化的植被都是被修剪得特别规矩的状态。类似于作品中的这样的大草卷都是出现在郊外的野生草场中，它们粗俗野旷，充满野生的体积感。在规矩的公共景观中很难看到这些大体量的野草的自然生长和凋零，它们也和我们看到的超级城市的外乡人一样，在卷曲的环境中生发，被管理者遗弃驱逐或者被修剪后镶嵌在城市人群中的某个角落。

Lawns and landscaped vegetation are all manicured in the city, while the wild Rolling Wind features a giant mass of grass which tends to be found only in suburban areas. It is difficult to see the grass naturally growing and withering in a well-regulated public view, similar to outlanders surviving in a twisted environment, abandoned or trimmed by urban regulators, and eventually turned into a mosaic in some corner among the crowds of city.

T8

展项名称：中国文人写意雕塑园（9件）
Exhibition Title: Chinese Scholars' Sculpture Park

策展人：方振宁
Curator: FANG Zhenning

参展人：吴为山
Exhibitor: WU Weishan

吴为山一直以挖掘和精研中国传统文化为人生命题，弘扬和传承中华传统文化。他长期致力于中国文化精神在中国雕塑创作中的融渗和表现，创作了大量具有影响力的雕塑，在世界多国展览并被重要博物馆收藏。他所创作的近五百件中国文化名人系列雕塑被季羡林等大师誉为"时代造像"，被国际评论界认为是"中国时代新精神的代表"。他的代表作《孔子》《孔子问道于老子》立于世界多个国家和驻外机构，如：巴黎中国文化中心、比利时中国文化中心、丹麦中国文化中心、新加坡中国文化中心、中国驻意大利使馆等。

WU Weishan has long been tapping into and drawing on rich traditional Chinese culture and committed to expressing the spirit of Chinese culture through his sculptures. He has crafted many influential works which are exhibited worldwide and collected by renowned museums. One of his sculpture collections, which consist of a host of historical figures of China, is hailed as an "Image Maker of the Time" by the prestigious scholar Ji Xianlin and regarded as "representing the spirit of a new China" by international critics. His most representative sculptures, "Confucius" and "Asking for the Way", can be found in various countries and overseas offices of China, including Chinese Cultural Centers in Paris, Brussels, Denmark, Singapore and the Chinese embassy in Italy.

《民族魂——鲁迅》/ 青铜 166x80x77cm / 2006

《齐白石》／青铜／高 1.93m／2012

《刘半农》／青铜 194x75x54 cm／2011

《曹雪芹像》／青铜 210x135x80cm／2010

《义勇军进行曲——聂耳》／青铜 220x120x175cm／2009

《悲欣交集——弘一法师》/ 青铜 183x77x46cm / 2006

《墨魂——黄宾虹》/ 青铜 186x70x68cm / 2006

《行走的人——作为哲学家的熊秉明》/ 青铜 163x70x40cm / 2006

《于右任》/ 青铜 220x65x50cm / 2011

T9

展项名称：内省腔
Exhibition Title: Introspective Cavity (exterior)

策展人：郭晓彦
Curator: GUO Xiaoyan

参展人：尹秀珍
Exhibitor: YIN Xiuzhen

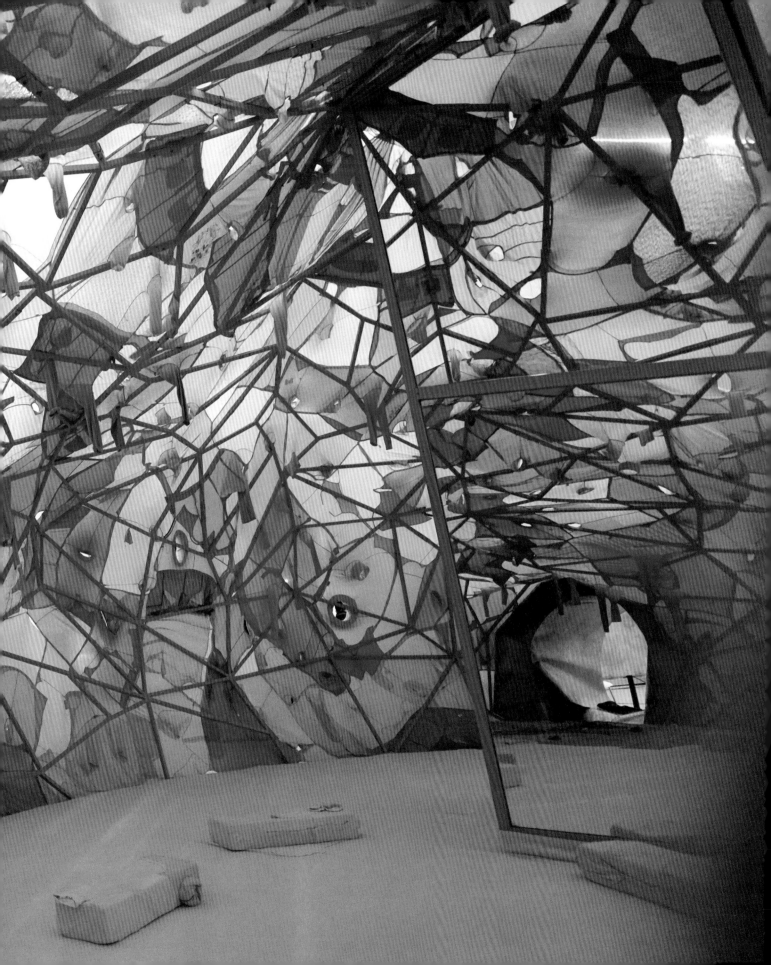

子宫是我们生命的起点，虽然我们曾经与她如此亲密，但我们没有任何视觉和感知记忆，而她一直存在于我们内心的深处。我们渴望回到母体，回到那个温暖、静谧、安详的状态。艺术家用收集来的不同人穿过的衣服创造一个孕育生命的空间，使我们有机会回到母体中。希望人们能回归这种状态，进行自我观照。在内心省察自我，"内省"是一种自我观察，是人对于自己的主观经验及其变化的观察；是对心理现象所遗留的"最初记忆"的观察；是对自己的思想或情感进行的考察；是对自己在受到控制的实验条件下进行的感觉和知觉经验所做的考察。现代社会中人很容易迷失，失去判断，变得茫然无从，不知道自己真正需要的是什么。回到原初，人们便受邀省察内心，探究个体的记忆、幻想和渴望。亦可暂时远离钢筋水泥、繁杂紧张的都市景观。当人们忙碌得如同机器人时，常常遗忘了自我。我们需要停下来，去休息一下。但我们又无法真正地回到原初，这个静谧幻觉之地仍然不时地被外界投射的不安所打扰，使这个满附不同人经历的器官空间，染了社会的属性。

The uterus is the starting point of our lives. Although we may have been so close to her, we don't have any visual and perceived memory of it. Eager to return to that matrix, we seek to return to that warm, quiet and serene state. The artist used clothes collected by different people to create a space for the birth of life, giving us an opportunity to return to inner world. The hope is so that people can return to this state, to reflect on themselves. "Introspection" is about self-observation, the observation of their own subjective experience and changes; it is the observation of psychological phenomenon left by the "initial memory"; it is the examination of their own thoughts or feelings; it is the inspection of the sensory and perceptual experience they undergo under controlled conditions. People too easily lose their judgment in modern societies. They don't know what they really need. Back to the start, people are invited to examine their hearts, explore individual memories, fantasies and desires. Getting away from the hustling urban world of concrete, if only for a while. When people become as busy as robots, they often forget true selves. We need to stop and take a break. But we can never really return to the start, since the quiet illusion is occasionally disturbed by the outer world. This biological space is intricately socialized with human experiences.

T10

展项名称：设计长椅
Exhibition Title: Design Bench

策展人：方振宁
Curator: FANG Zhenning

参展人：靳烈
Exhibitor: JIN Lie

参展的六件作品,是为配合上海SUSAS公众艺术的特性而特别设计的。作品以可以实际使用的长椅的形式出现。六条长椅以结构主义艺术造型为出发点,融合艺术、公众空间实用品于一体,通过实际使用性缩小艺术与观众的距离,是公众空间、艺术与观众的连接点。

The six benches are specifically designed for SUSAS 2017 and its call for art in public space. They are meant to be practical. The basic idea comes from constructivism whereby utility and art become one. The distance between art and people is reduced and they consequently become a hub for exchange and communication between public space, art and the viewers.

T11

展项名称：风律
Exhibition Title: Rhythm of Wind

策展人：李翔宁
Curator: LI Xiangning

参展人：盛姗姗
Exhibitor: SHENG Shanshan

《风律》是利用回收的金属材料创作的大地艺术作品,矗立在广阔的原野中,形成一组绵延错落的创作。作品在形式上通过将钢材精心排列,使之在空间中呈现出如同在风作用下蜿蜒的姿态,表现了一种风的律动感。这些日常熟悉的钢架结构,在被抛却了功能性之后,呈现的是其作为艺术品而不同于日常的形态,这些冰冷属性的材质仿佛被赋予了生命的内在,在风的洗礼下,在自然中重生。观者能够在冰冷钝重的机械和轻盈的风的对比之下,在自然和人为创造的风景的巧妙结合之间,从不同角度看到作品在自然中的张力,感受到生命的内在力量和自然诗意之间的相互结合、延伸和拓展。在夜晚,借助太阳能灯的照明效果,旧钢材斑驳的肌理与光的结合在夜晚展现出一种独特的神秘感。观者在不同的光线环境下,亦会随之产生不同的联想。

Rhythm of Wind is a Land Art piece utilizing repurposed metalic material. The rods are erected in a curvilinear formation in the environment, suggesting a flow and cadence of wind patterns. The work seeks to imbue characteristics of nature into man-made materials, as a reminder of our place in the larger scope of the environment. Urban construction, nature and people who live in this land make this connection with *Rhythm of Wind* to create a sense of place.

T12

展项名称：仓声·品
Exhibition Title: Cang Sheng · Pin

策展人：李翔宁
Curator: LI Xiangning

参展人：苏丹，王宁，张荐
Exhibitors: SU Dan, WANG Ning, ZHANG Jian

一千个太阳能自发声的音箱和一千个不同的器皿，以及特殊的空间组合而成一个具有感染力的声场。场所的特质和食用道具的组合打消了音乐的模糊性，它们相互引导构建新的寓意。只要有光，声音的叙事就开始了，过去的粮仓也就转化成为一个精神食粮的发生器。

One thousand solar energy sound players, one thousand different vessels and a unique space combine to form an acoustic field that seems all too contagious. By combining characteristic space and vessels to break with the vagueness of music, they lead each other to create new meanings. The sound begins to tell its story with light, and this space, previously used as a granary, will be transformed into an acoustic machine of spirits.

T13

展项名称：2340 洞
Exhibition Title: 2340 Holes

策展人：李翔宁
Curator: LI Xiangning

参展人：于幸泽
Exhibitor: YU Xingze

人类文明的初光发自洞穴，人探索天地奥秘亦是管中窥豹、洞察玄理。人的大脑就像一个微观宇宙，其复杂之谜至今未解，有着无限的潜能与可能，想象力与创造力无法估量。每层3根柱子象征天、地、人，它们层叠交错与时间共同构建起一个四维空间，而39根柱子共有2340个孔洞，与柱子间的无数空隙交织成一个近乎无限的多重关联的空间系统，暗示着时刻都在进行着多维几何级演变的未知世界。

Human civilization originated from caves, and all humans can expect in their exploration is a fragment of the universal mystery. The brain, like a micro universe with infinite possibilities, has immeasurable imagination, creativity and complexity yet to fathom. Each of the thirteen layers of 2340 Holes has three columns, symbolizing the universe, Earth and Terrance, constituting a spatial-temporal field. Thirty-nine columns have 2,340 holes in all. The holes and spaces in between make up a complex structure, implying an unknown multidimensional world which continuously changes in its geometric series.

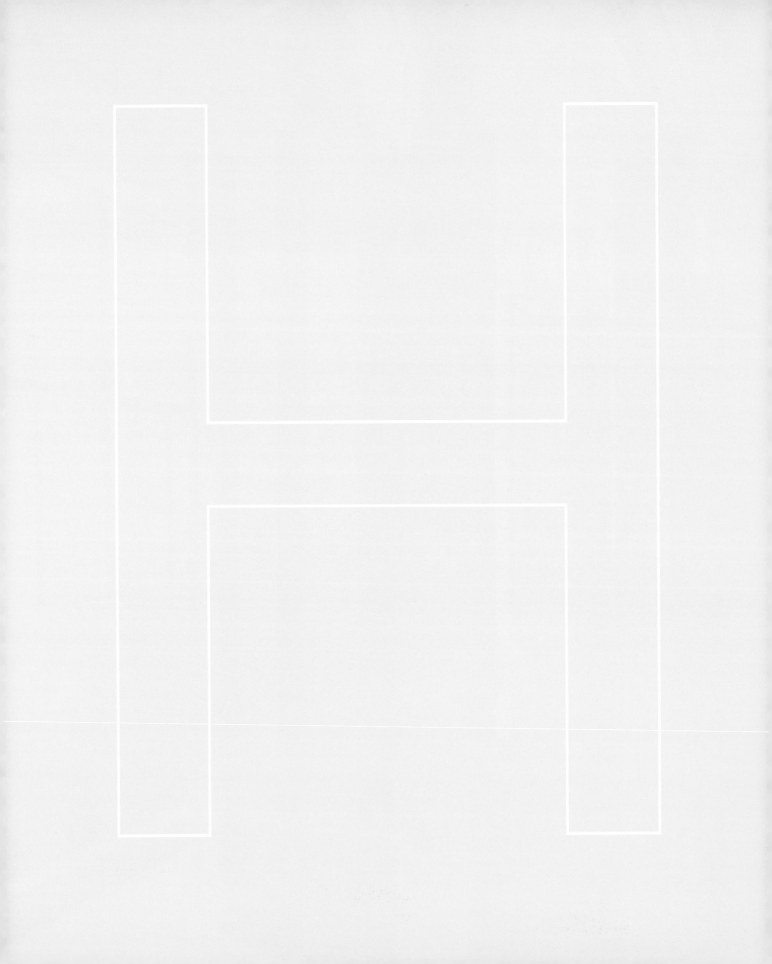

城市空间的特征与其所处社会文化环境密切相关。无论是全球城市还是地方城市，它们的城市空间都因为其独特的社会文化基因而具有吸引力。前者得益于多元文化的混合和相互激发，后者则依赖于独特的地方资源而具有识别性。我们通过呈现多样的地域、历史、机制和价值观，把整个展区转变成内含多样社会文化的场所，激发上海在全球 - 地方城市建构过程中的积极思考。

Characteristics of urban space are closely related to the social and cultural environment. Urban spaces of metropolises or towns are both attractive for their unique sociocultural facets. The former benefit from various cultures mixed up and inspiring each other while the latter rely on unique local resources. By showing diversified geographies, histories, institutions and values we transformed the entire exhibition area into a place embracing a variety of social and cultural contexts, and inspire positive thinking of the global-local construction process in Shanghai.

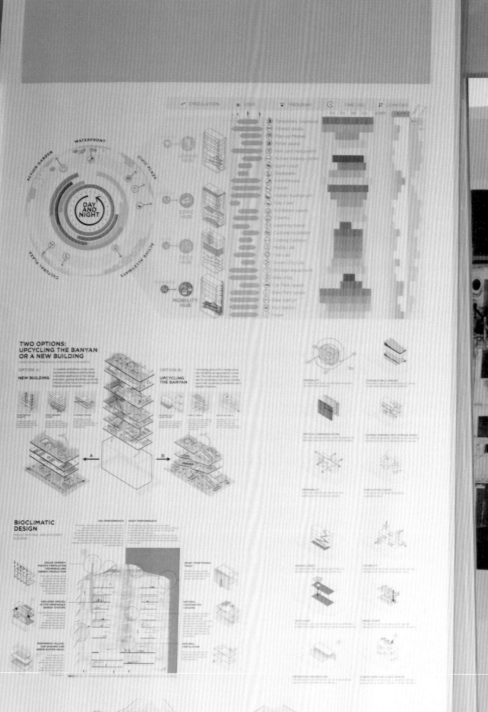

H1

展项名称：回音：建筑与社会
Exhibition Title: The Echo from Society: Architecture and Contemporary Challenges out of Established Agendas

策展人：冈萨雷斯，田唯佳
Curators: Placido Gonzalez Martinez, TIAN Weijia

参展人：弗朗西斯·克雷，荷西·拉艾多，安德烈斯·哈克，王子耕，生态系统城市研究室，乡村城市共建工作室，白德拉研究所
Exhibitors: Francis Kéré, Jorge Raedó, Andrés Jaque, WANG Zigeng, Ecosistema Urbano, Rural Urban Framework, Instituto Pedra

在工业革命之后现代主义运动开始时,建筑与社会的关系便开始持续地被讨论。从那时起,建筑表现得像是一个能够为个体提供庇护的工具,支撑着人类勇敢地去面对广阔无边的大都市。社会与社区的矛盾问题在19世纪末20世纪初期首先被德国的社会学者如滕尼斯、西美尔等提出,之后这样的讨论自亚当斯开始融入美国的主流。特别是在芝加哥,由于帕克和沃斯的影响,这个话题在那里埋下了深深的根基。20世纪末21世纪初,对社会与建筑问题的讨论变得越来越复杂。伴随着后殖民时期出现的关于性别、民族、年龄等问题,著名学者如卡斯特尔斯、萨森、巴巴哈等都在进行相关的讨论和研究。建筑与社会所涉及的话题非常广泛,这也是当代社会所面对挑战的最大特点,但是它们全部指向了建筑这个学科所面对的终极问题,那便是空间的生产。"回音"这个标题的确定正是为了回应本次展览的主题"thisCONNECTION",借用建筑、多媒体、图像等手段来反映社会问题与空间生产的联系。这个板块展示的作品分别来自七位建筑师、学者及机构,他们的实践与研究都重点关注本话题的相关领域,他们分别为:弗朗西斯·克雷、荷西·拉艾多、安德烈斯·哈克、王子耕、生态系统城市研究室、乡村城市共建工作室、白德拉研究所。

The relationship between architecture and society has been recurrent since the Industrial Revolution. Architecture has since then offered a framework for the public realm, while at the same time provided protection, safety and identity to its denizens who confront the metropolis immensity. The dialectic between society and community, originally defined by Ferdinand Tönnies, Georg Simmel, Robert E. Park and Louis Wirth among others, has evolved into a complex debate in the 21st century, imbricated with postcolonial discourse as well as that of gender, race and age. Manuel Castells, Saskia Sassen, and Homi K. Bhabha are among those offering their insights. The relationship between architecture and society is an encompassing issue and fundamental challenge of our era, while in the end, it's about what architecture has been about: creating spaces. The selection incorporates the works of architects and urban planners currently engaged in this production, building a narrative that explores a response – an echo to the current needs of local communities in the framework of global societies. Included are works by Francis Kéré, Ecosistema Urbano, Jorge Raedó, Rural Urban Framework, Andrés Jaque, WANg Zigeng and Instituto Pedra.

H2

展项名称：南京长江大桥记忆计划
Exhibition Title: Memory Project of the Nanjing Yangtze River Bridge

策展人：鲁安东
Curator: LU Andong

参展人：LanD Studio
Exhibitor: LanD Studio

盒子向公众展示大桥记忆，记忆的载体可以是一张旧照、一件旧物、一份历史资料、一个替换下来的大桥部件，等等。这些记忆数据同时存储于网络，盒子上有对应的二维码。参观者通过移动终端扫描二维码会听到一段回忆，亦或是看到一段影像，甚至是更多人对大桥的寄语。

存储有大桥记忆的盒子，通过巧妙设计，在展馆内叠加成立体的巨型二维码，同时也是一件立体山水装置作品，呈现出集体的、由每个人的记忆组成的景观。就像当年的大桥是由聚力而成，这个景观将再次见证人与人联合起来的伟大力量。

Memory boxes are displayed as an installation about recollections. It seeks to display the memory of the Nanjing Yangtze River Bridge to the public. Carriers of memory may be an old picture, an old utensil, a fragment of someone's voice, historical data, a component of the Bridge or some art work. This documentary is also placed on the Internet so that visitors can scan the QR code on each box and know a piece of memory, see a video clip, or see how others feel about the Bridge.

Meanwhile, the memory boxes are designed into a landscape installation in the pavilion, presenting both collective and individual memories. Just as the Bridge was formed by force of cohesion, the landscape will witness the profundity of human union once again.

RED FURURISM
TECHNOLOGICAL MONUMENTALITY IN CHINESE ARCHITECTURE THROUGH CASE STUDY OF THE NANJING YANGTZE RIVER BRIDGE
红色未来主义——南京长江大桥与中国建筑中的技术纪念碑性

H3

展项名称：濑户内国际艺术祭
Exhibition Title: Setouchi Triennale

策展人：北川富朗
Curator: Fram Kitagawa

参展人：林舜龙
Exhibitor: LIN Shunlong

瀬户内国际艺术祭为一个每三年举办一次的当代艺术节，于2010年首次举办，由北川富朗担任总监。2016年的第三届以"海的复权"为主题，于濑户内海的12座岛屿、高松港及宇野港举办。在现今全球化的世界中，在同质化和效率化的趋势之下，这些岛屿也因人口老化和衰落而失去其独特的特征。濑户内三年展的目的，是为了让岛屿恢复活力，力图使濑户内海成为世界各地的"希望之海"。林舜龙的《跨越国境·潮》在2016年濑户内三年展的小豆岛海岸旁展出。作品里包括196位用沙子做成的儿童，代表日本认可的国家数量。风吹日晒之下，人像慢慢瓦解、归于沙砾，好似因成人世界或交战国家的残酷蹂躏，而在世界上流浪寻找家园的弱势儿童。我们也在这个展位的中心，将此沙雕复制展出。

Setouchi Triennale is a contemporary art festival held every three years. It was first held in 2010 and the next Triennale will take place in 2019. Fram Kitagawa has been the general director of the Triennale. There are also various activities called Art Setouchi outside the period of the Triennale. The 3rd Triennale in 2016 continued to pursue the theme of Restoration of the Sea. It was held over three sessions, spring, summer and autumn, for a total of 108 days on the 12 islands of the Seto Inland Sea, as well as in the ports of Takamatsu and Uno. In today's world, with increasing homogenization and streamlining, the islands were losing their unique characteristics because of an aging and decreasing population. The Setouchi Triennale is held with the aim of returning buoyancy to the islands, and seeks to make the Seto a "Sea of hope" for all regions of the world. The installation "Beyond the Border- Tide" by LIN Shunlong were exhibited along the beach of Shodoshima at the Setouchi Triennale 2016. A total of 196 sand sculptures of children represent the number of countries recognized by Japan. Exposed to the sun, wind and rain, the figures slowly disintegrated, returning to sand, a symbol of vulnerable children who wander in search of a home or shelter, tossed about by the cruelty of adults and warring nations. Now the replicas are placed in the center of pavilion.

H4

展项名称：连接：空间移动
Exhibition Title: Connection: Space Movement

策展人：郭晓彦
Curator: GUO Xiaoyan

参展人：刘韡，沈远，邱志杰，徐震，黄永砅，何岸，王郁洋，奥拉维尔·埃利亚松，刘建华，程然，汉斯·欧普·德·贝克，多米尼克·冈萨雷斯·福斯特，法比安·纪劳德
Exhibitors: LIU Wei, SHEN Yuan, QIU Zhijie, XU Zhen, HUANG Yongping, HE An, WANG Yuyang, Olafur Eliasson, LIU Jianhua, CHENG Ran, Hans op de Beeck, Dominique Gonzalez-Foerster, Fabien Giraud

"连接：空间移动"展览集合了15位中外优秀艺术家的作品，力求多视角、多元化地审视艺术与城市及空间错综复杂的关系，构建一个临时性的叙事。艺术家的作品以录像、摄影、装置、概念艺术与雕塑等形式与此刻的空间发生着连接。

这些创作是艺术家对于自身所处环境的思考，抑或是对于现状、问题提出的深刻质疑，均与空间发生着密切的关系——艺术家将社会感知与信息反馈汇聚起来，编织出似曾相识却又悬疑陌生的空间叙述，或者另一种纬度的构造，非取材于任何客观形状，而是由言语和时间搭建的叙事体，以特有的方式定义或是重新定义着特定空间与周围世界的多重连接。

展览集合了丰富多样的空间叙述形式，以艺术家独有的视角审视我们所在的世界。参展艺术家以城市的社会文化为起点，探寻、反思了他们所处世界的境遇，带来多重文化与言语模式的艺术体验。

Artwork from fifteen prominent Chinese and foreign artists comes together for the exhibition "Connection: Space Movement", which has managed to Interpret from multiple viewpoints the intricate relationships between art, city and space and make a temporary narration. These works are connected to the present space with videos, photography, devices, concept art, sculpture and more.

These works, which incorporate thoughts about the artists' circumstances or doubts about current issues, have an intimate relationship with space, expressing an intense experience and appetite to discuss existential issues. The artists gather indescribable perceptions and feedbacks of the society in an effort to rebuild this mental space, itself a fantastic coagulation, to create a narration in space which seems to be faintly familiar as well as alien. Otherwise, the artwork is a construction of an other-dimension a descriptive style set up by time, not based on any objective shape. The artwork takes on a multimodal dialogue with space.

"Connection: Space Movement" gathers varieties of spatial narration and interpreted our world form the artists' unique angles of view. Participating artists take the cities' social culture as the starting point, exploring and rethinking circumstances around the world and bringing up artistic experience for multiple cultural and discursive patterns.

展项名称：迷中迷 / 参展人：刘韡
Exhibition Title: Enigma / Exhibitor: LIU Wei

展项名称：金属的语言 / 参展人：徐震
Exhibition Title: Metal Language / Exhibitor: XU Zhen

展项名称：你的偶遇 / 参展人：奥拉维尔·埃利亚松
Exhibition Title: Your Chance Encounter / Exhibitor: Olafur Eliasson

展项名称:狂人日记 / 参展人:程然
Exhibition Title: Diary of a Madman / Exhibitor: CHENG Ran

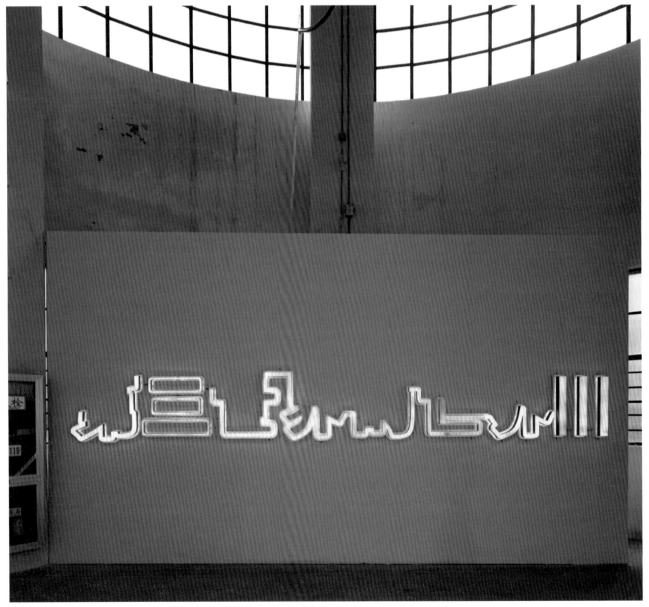

展项名称：第三世界 / 参展人：何岸
Exhibition Title: The Third World / Exhibitor: HE An

展项名称：2号双翼 / 参展人：黄永砯
Exhibition Title: Double Wing / Exhibitor: HUANG Yongping

展项名称：容器 / 参展人：刘建华
Exhibition Title: Container / Exhibitor: LIU Jianhua

展项名称：水中倒影 / 参展人：刘建华
Exhibition Title: Shadow in the Water / Exhibitor: LIU Jianhua

展项名称：《编制系列》/ 参展人：邱志杰

Exhibition Title: Spree, Trailing Light, Regression, Draft, Stalactite, Sculpting in Time, Book Burning Edict / Exhibitor: QIU Zhijie

展项名称：无墙 / 参展人：沈远
Exhibition Title: Without Wall / Exhibitor: SHEN Yuan

展项名称：无题 / 参展人：王郁洋
Exhibition Title: Untitled / Exhibitor: WANG Yuyang

展项名称：决心 / 参展人：汉斯·欧普·德·贝克
Exhibition Title: Determination / Exhibitor: Hans op de Beeck

展项名称：RIYO / 参展人：多米尼克·冈萨雷斯-福斯特
Exhibition Title: RIYO / Exhibitor: Dominique Gonzalez-Foerster

展项名称：直边 / 参展人：法比安·纪劳德
Exhibition Title: Straight Edge / Exhibitor: Fabien Giraud

H5

展项名称：空间的边界
Exhibition Title: The Frontier of Space

策展人：方振宁
Curator: FANG Zhenning

参展人：王昀
Exhibitor: WANG Yun

该作品为参展建筑师王昀多年来对于空间的操作、空间领域观念的拓展及实体设计的集中呈现，对于空间的边界进行了全方位探索。展览由7个部分构成，内容包括：音乐与建筑、书法与建筑、聚落与建筑、园林与建筑、斗拱与建筑、废物与建筑、自然与建筑。另外还包括21个空间建筑的实际项目：60平方米极小城市、善美办公楼门厅增建、百子湾小区幼儿园、百子湾小区中学、庐师山庄会所、庐师山庄A住宅、庐师山庄B住宅、石景山财政培训中心、西溪湿地—方体空间、西溪湿地—梅花、西溪湿地—喇叭、西溪湿地—烟囱、西溪湿地—椭圆住宅、西溪湿地—桥宅、西溪湿地—长宅、西溪湿地—长方体、西溪湿地—水上漂、西溪湿地—框景舞台、萨蒂的家、吴家场幼儿园、吴家场A1等。这些项目展示了建筑师本人对于设计的思考。如此，不同领域的连接结果就构成了空间的边界。

The Frontier of Space comprehensively demonstrates the architectural designs of Wang Yun and how his conceptualization of manipulating, augmenting space has evolved. The exhibition consists of seven sections, including Music and Architecture, Calligraphy and Architecture, Settlement and Architecture, garden and Architecture, Tou-Kung and Architecture, and Architecture, Junk and Architecture, Nature and Architecture. And actual projects of 21 space buildings, including: a small city of 60 square meters, ShanMei Office Building Foyer, Baiziwan Kindergarten, Baiziwan Middle School, Lushi Mountain Villa, the Clubhouse of Lushi Mountain Villa, the Shijingshan Bureau of Finance Training Center, Xixi Wetland-cube space, Xixi Wetland-plum blossom, Xixi Wetland-trumpet, Xixi Wetland- chimney, Xixi Wetland-oval residence, Xixi Wetland-bridge house, Xixi Wetland-long house, Xixi Wetland- cuboid, Xixi Wetland-water float, Xixi Wetland-frame stage, Erik Satie´s home, kindergarten of Wujiachang, A1 apartment of Wujiachang, showing the architect's thoughts about the design. These connections between different areas naturally give rise to a spatial frontier.

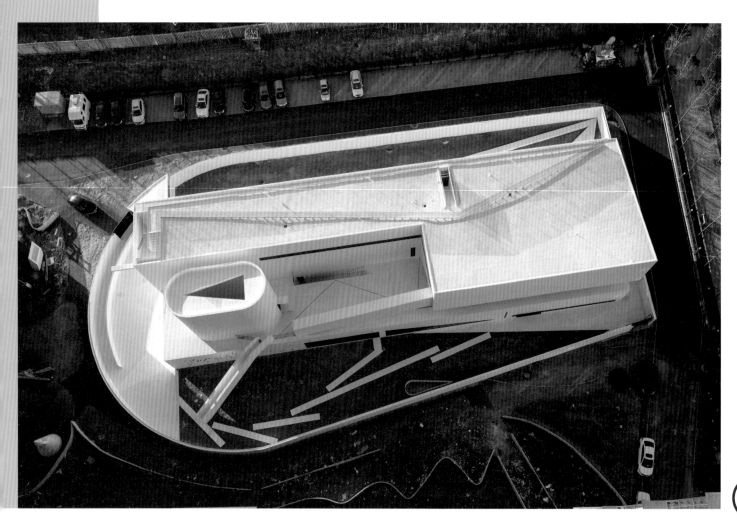

我们生活在这个世界，常常分不清什么是虚构，什么是现实。黄莺着迷于虚拟与现实之间的灰色地带。虚拟产生于现实的挤压，将虚拟形态转换为一种可以触摸和感知的实体，延伸到现实环境之中，它将一个完整的体系打开了一个缺角，用不同的和相同的特质交融，产生一个新的共联的虚拟现实的空间，影响未来。《你是如此温柔》数字绘画来自于艺术家将自身旅行的记忆图像通过电脑程序解码和编码，《幻想速度》投影装置是艺术家通过对物理速度身临其境的感知和对碎片化网络时代的反思，探寻真实与虚拟、碎片与共生、记忆与遗忘、加速与疏离等相互粘连的错综关系。两件作品构成的整体装置将个体和公共的记忆与未来连接在一起，形成一个沟通历史与想象的场域。

We often have difficulties in distinguishing between fiction and reality in this complex world. The artist is fascinated by the blurry terrain between these two worlds. We can transform virtual forms into tangibles that can be touched, perceived and extended as part of the real world. These kinds of activities serve as interference that leads to breakthroughs. A new united virtual- reality space will come out with the same and different characteristics and further influence the future. To decode and encode photos taken by the artist in her journeys with software, she attempts to search for subtle and indescribable connections between memory, perception and the environment in digital paintings titled "You are so Tender". The "Fantasy Speed" installation has its inspiration from a feeling of speed in the driving experience, which the artist paraphrased and recoded it digitally. The artist intends to explore relationships on a deeper level such as the real-virtual, fragment-mutualsm, remembering-forgetting, acceleration- alienation. The installations connect both individual and collective memory with the future, forming a field of communication between history and imagination.

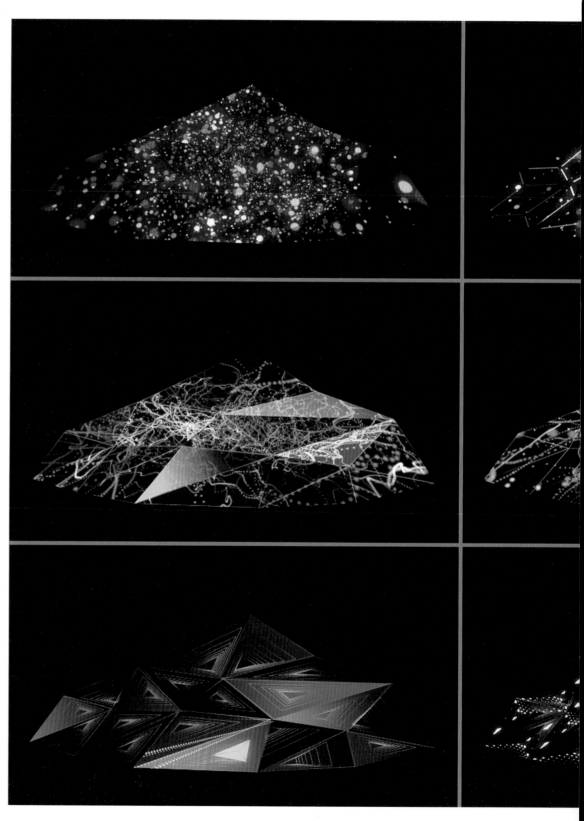

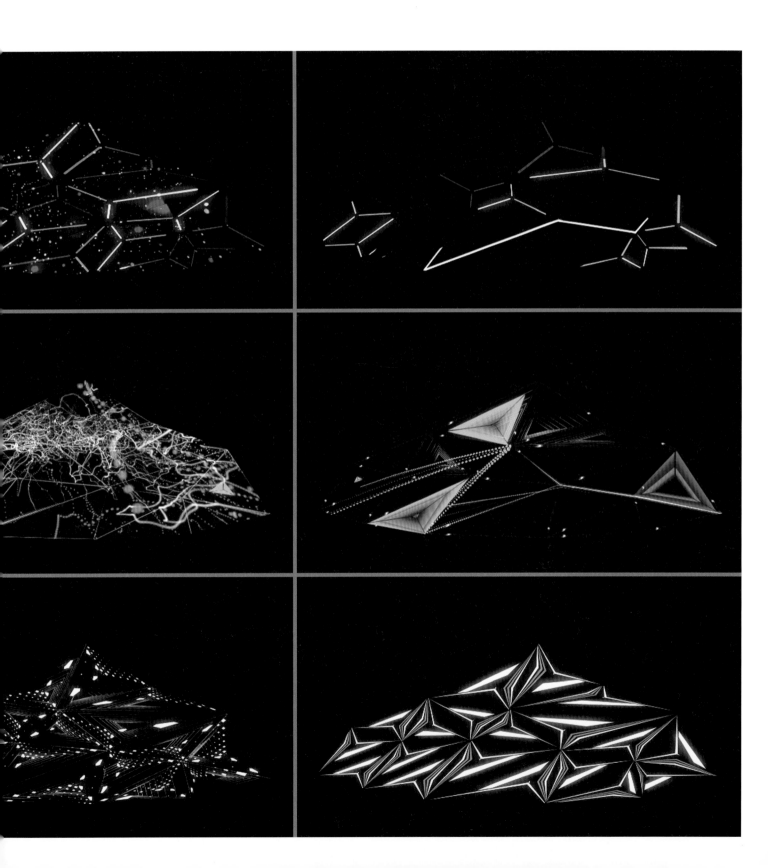

H5

展项名称：寻找马列维奇
Exhibition Title: Malevich Finders

策展人：方振宁
Curator: FANG Zhenning

参展人：张永和
Exhibitor: Yung Ho Chang

这一组六个装置，都由可旋转底座和各个不同的取景窗口组成，故名取景器，由于在空间中有了寻找几何图形的目标，故可更名取形器。

每个取形器分别对应一个几何图形，其中五个对应环境，分布在柱子和筒仓上；一个在对应的互动者本身，在一顶可供观展人佩戴的帽顶上。

取形器与图形之间并不是简单的对应：取形器的位置和视角会使被观察的图像产生形变，因此我们对几何图形进行了调整，比如通过方形取景框仰视，会发现和柱上的一个梯形对应了；又由于远距离透视，通过长条形取景框会发现与延展到三根柱子上的长条形对应了。在其他取形器上均可以得到类似体验。同时，观展者在搜寻过程中有机会对场地产生更深入的甚至特殊的认知。

Six devices, composed of rotatable bases and distinct apertures to capture views, are named View-Finders. Since there are targets of geometric figures to be viewed in the space, these can also be called Form-Finders.

Each Form-Finder corresponds to a geometric figure. Five of them represent the notion of environment and are placed on columns and the silo. The last one refers to the viewer, attached to a cap he or she wears.

The correspondence between the Form-Finders and geometric figures is not straightforward. The image of the geometric figures can be distorted due to the position and viewing angle of the Form-Finders. So we have adjusted the geometric figures. For instance, by looking up through a square-shaped Form-Finder would only find it to coincide with a trapezoid- shaped target. Another example would be due to long distance perspective, viewing through a bar shaped Form- Finder would feel like that the aperture coincides with a bar shape that is extended onto three columns. Similar experiences can be drawn from other Form-Finders. At the same time, while looking through the devices, the viewers can perceive a deeper understanding of the site.

展项名称：天路
Exhibition Title: Cosmos

策展人：方振宁
Curator: FANG Zhenning

参展人：李磊
Exhibitor: LI Lei

都没有根
狂热的焦虑
洒在
任何一点潮湿的地方
上帝的网
打不尽拥挤的欲望
激情
撞击出缠绵
在所有的夜和白天
报以烟花
和彩屑
满天的
和满地的
都没有根

All rootless
fanatical anxiety
is scattered over
wherever it's damp.
God´s web
can't capture all fulfilling desires.
Passion
generates tenderness in strikes.
Day and night,
it sets off fireworks
and throws confetti back,
filling the whole sky,
covering up the boundless earth,
all rootless.

H5

展项名称：蝴蝶夫人
Exhibition Title: Madame Butterfly

策展人：方振宁
Curator: FANG Zhenning

参展人：吕越
Exhibitor: Lyu Yue (Aluna)

作品采用印有蝴蝶的蓝印花布与布料刻成的蝴蝶进行互动，将蓬裙的弧形与鸟笼吻合，把厨娘的围裙和大摆礼服裙进行不协调拼接，加上不同年代的旗袍，创造出淑女与厨娘、自由与禁锢、中国土布与西式裙撑、手工印染与激光雕刻、平面与立体、过去与现在、阴与阳、虚与实，那些看似不相干的东西似乎又显现了相互支撑的和谐。矛与盾共存，正是创作者要表达的内容。

In this work, the fashion artist arranges printed blue calico and fabric-made butterflies in a manner that they are interactively entangled and creates a silhouette of bouffant gown like a birdcage. Meanwhile, the artist creatively matches a female chef's apron with a ball gown. The work is accompanied by several version of Qipao dresses from different eras. One finds various polar pairs in the work: masters and servants; lady and female chef; freedom and imprisonment; Chinese hand-woven cloth and Western crinoline; hand dyeing and laser engraving; planar and stereoscopic; past and present; yin and yang; fiction and reality. Here, these apparent opposites show a harmony. This coexistence between opposites is the main theme that the artist wishes to express.

H5

展项名称：小库：人工智能时代的未来都市
Exhibition Title: Future Metropolitan of AI Era

策展人：方振宁
Curator: FANG Zhenning

参展人：何宛余，小库科技
Exhibitor: HE Wanyu, XKool

第四次工业革命热烈上演,人类站在了奇点边缘,社会正经历超速发展和变化。在这一次革命中出现了非常多足以改变整个世界面貌并日趋成熟的技术,包括人工智能、大数据、区块链、云计算、虚拟现实、增强现实、无人机、无人驾驶、机器人……它们都代表了在每个不同领域的巨大想象空间和未来机遇。然而,我们的城市是否做好了迎接新技术的准备?我们希望通过展示新技术在城市日常中的运用,将人工智能的设计生成和增强现实的大数据叠加在城中村场景中呈现。展现未来科技渗透城市的各个角落,甚至体现科技加持后原本"落后"的地区如何在其他维度上超越城市"优质"区域。我们希望通过此次展览展示这些技术的未来图景和实际可行性。

As the Fourth Industrial Revolution is going on, mankind has come around the verge of singularity, while the society is undergoing rapid changes. New technologies such as AI, Big Data, Block Chain, Cloud Computing, VR, AR, Drones, Auto piloting and Robots are arising and maturing. They represent borderless imagination and opportunities. Is our urban space ready to embrace them? We want to apply AI intelligent design, AR and Big Data on scenarios of "urban villages", introducing technologies to daily urban life, and revealing infiltration of new technologies into every corner of city. Furthermore, we want to propose how technologies will help underdeveloped areas to catch up across different dimensions. By presenting our ideas and innovations in this exhibition, we demonstrate our projection of the future and feasibility of new technologies.

H6

展项名称：内心（影像）
Exhibition Title: Inner Space

策展人：方振宁
Curator: FANG Zhenning

参展人：沈伟，马岩松
Exhibitors: SHEN Wei, MA Yansong

在某种程度上,建筑和舞蹈,都是用来表达的语言。

建筑师马岩松和舞蹈家沈伟分别作为这两个领域的代表人物,用各自拿手的语言,相互融合、交锋以及探索,最后形成了6分50秒的短片。建筑师马岩松作为这支短片的创作者之一毫不奇怪,因为短片探寻的建筑,正是他的MAD建筑事务所设计的城市新地标建筑——哈尔滨大剧院。

哈尔滨大剧院于2015年建成。大剧院与自然、城市有机互动及融合,就像从自然中生长而立,成为一处人文、艺术、自然相互融合的大地景观。这座位于东北的文化地标,建成当年便引起全球持续关注,先后被誉为"年度文化建筑""全球最佳音乐厅",荣获"最佳表演空间奖"称谓,CNN更称其为"超越悉尼歌剧院的艺术作品"。

建筑,可被看作是城市中的艺术品。舞者与这座属于音乐艺术的建筑的互动——舞者因空间感受而作的舞蹈创作,空间因舞者的演绎而更鲜明的个性色彩,激活了艺术的多元表达性。敏锐的观众能在捕捉肢体的伸展与流动时,感受到来自舞者身体张力和空间体的相互交锋。它试图呈现出一种想象力,超越肢体的局限,也超脱空间的轮廓。

To some extent, architecture and dance are both expressive languages.

As top figures in their own professions, Ma Yansong the architect and Shen Wei the dancer have co-produced a brief video titled "INNERSPACE" where their talents met, collided and resulted in an explorative experience.

The harbin Opera House completed in 2015 was designed as a response to the force and spirit of this northern city's untamed wilderness and frigid climate. It is a cultural center for the future—a tremendous performance venue, as well as a dramatic public space that embodies the integration of humans, art and city identity, while blending with the natural environment. Since the harbin Opera house was made public, it has been awarded "Building of the year—Culture", "the world's best concert halls", "the best performing space". CNN reported it with the title "Move Over, Sydney".

Buildings can be seen as a city's Most prominent art pieces. The interaction between the dancer and the opera house — choreography inspired by architecture, and the more distinct identity of the architecture through presentation of dancer---- further activates its diverse characters. Through the dancer's stretching and fluid movement, the video clip gives the audience a feel for the interplay between the dancer's tension and the surrounding space. It expresses a new way of imagination, exceeding the limitation of bodies, as well as the outline of space.

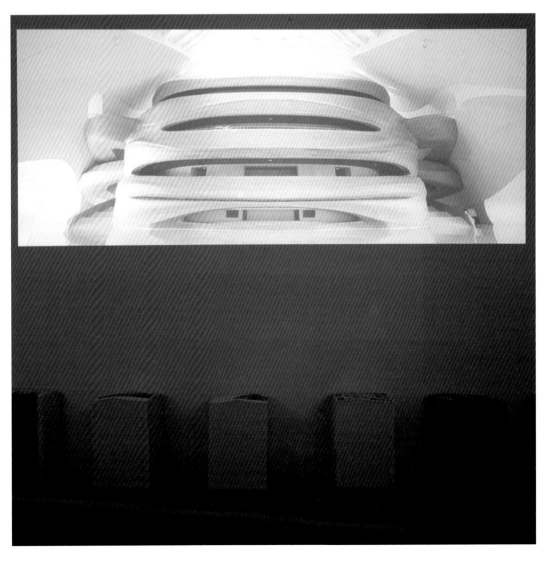

基础设施连接
INFRASTRUCTURE

新的技术设施不仅是物理的,也是基于项目的。对于传统道路和桥梁的新的利用方式,以及新的交通网络和互联网技术,都给空间和时间的连接带来了新的体验。除此之外,我们还将展示:当代建筑学对于空间连接的思考已经超越了传统的基础设施模式,这种超越不仅限于形式本身的革新,更在于对公共和社群活动以及人们身体体验的关注。建筑学已经成为城市文化空间生产的基础设施。

New technological facilities are project-based physical entities. New ways to utilize traditional roadways and bridges, along with a new transportation network and Internet-based technologies, have brought about fresh experiences in connecting space and time. Besides these novelties, we demonstrate how today's architecture has surpassed traditional modes of infrastructure concerning understandings of spatial connections. Such superiority lies not only in scientific innovation itself, but also in attention paid to the public and community activities, as well as to people's physical experience. Architecture is now a literally infrastructure of how culture arises from urban spaces.

展项名称：桥
Exhibition Title: Bridge

策展人：李翔宁，姚微微，谢雨晴
Curators: LI Xiangning, YAO Weiwei, XIE Yuqing

参展人：大舍建筑事务所，无止桥慈善基金，非常建筑事务所，赫斯维克事务所，JCFO，马德里欧洲大学
Exhibitors: Atelier Deshaus, Wu Zhi Qiao (Bridge to China)Charitable Foundation, Atelier FCJZ, Heatherwick studio, James Corner Field Operations, Universidad Europea de Madrid

桥,是物理空间连接的基本类型。

建筑师的桥,不仅是基础设施连接最基本的体现,还融合了建筑师对于城市公共空间、建筑、景观和艺术的多重思考。建筑师利用多功能复合的桥,提供让人驻足的公共空间;通过自成景观的设计,更加激发空间活力。桥梁结构不仅是建造的基础,同时也成为融功能和形态于一体的表现形式。不同的地域环境赋予桥不同的社会意义与场地特质,使桥以一种锚固的姿态承载着川流不息的人群,让穿越和停留并置,架构着一段更丰富的场所记忆。

桥的专题展,旨在更广阔的语境下呈现建筑师关于桥的多元思考,深入探讨基础设施建设对于塑造积极的城市公共空间所具备的多样可能性——从具体地点的连接,扩大到公共空间的聚集,再到和社会文化的对话,在连接的实验中,桥突破其固有的形式与内涵,跟随着不可见的鲜活的思维延伸。

Bridges are a basic structure connecting physical spaces.

Bridges are not only most basic embodiment of spatial connection, but also incorporate the architect's multi-level thinking about a city's public space, architecture, landscape and art. Architects design bridges with multiple functions to provide people with a public space to stay in, and as a landscape to invigorate the space. Structures are not only the basis of construction, but also a way to express functions and forms. Geographies bestow social significance and characteristics on bridges, which persistently bear passenger flows as a memorable place for passing and staying.

This special exhibition aims to showcase architects' diverse thinking about bridges in a broader context, and explore in-depth possibilities of infrastructure construction in positively shaping urban public spaces. In this connection experiment, from specific connections made to public convergence to social-cultural dialogue, bridges have broken through its inherent form and connotation and extend in line with an invisible yet vivid mindset.

展项名称：日晖港桥
Exhibition Title: Footbridge on Rihui River

参展人：大舍
Exhibitor: Atelier Deshaus

展项名称：贵州谢家村无止桥
Exhibition Title: Wuzhi Qiao, Xiejia Village, Guizhou Province

参展人：西安交通大学、西安建筑科技大学、香港中文大学
Exhibitors: Xi'an Jiaotong University, Xi'an University of Architecture and Technology, The Chinese University of Hong Kong

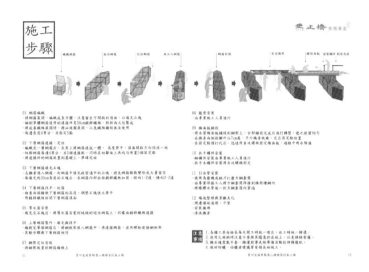

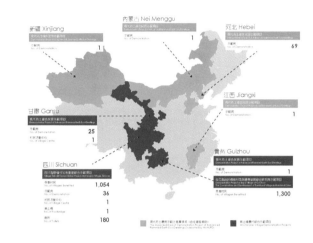

示範項目分佈及推廣領域（2007年4月至2017年8月）
Location of Demonstration Projects and Impacted Areas (2007.4 to 2017.8)

展项名称：吉首美术馆桥
Exhibition Title: Jishou Art Museum Footbridge

参展人：非常建筑
Exhibitor: Atelier FCJZ

展项名称：滴水湖景观七桥
Exhibition Title: Dishui Lake Green Belt 7 Bridges

参展人：白德龙
Exhibitor: Pedro Pablo Arroyo Alba

展项名称：卷桥
Exhibition Title: Rolling Bridge

参展人：斯维克事务所
Exhibitor: Heatherwick Studio

展项名称：花园桥
Exhibition Title: Garden Bridge

参展人：斯维克事务所
Exhibitor: Heatherwick Studio

展项名称：高线公园
Exhibition Title: High Line Park

参展人：JCFO
Exhibitor: James Corner Field Operation

12

展项名称：物联生产
Exhibition Title: Interconnected in Production

策展人：一造科技
Curator: Fab-Union

参展人：一造科技团队，冶是建筑工作室
Exhibitors: Fab-Union, YeArch Studio

"物联生产"是一份针对中国当下生产状况的研究报告,通过对长三角地区生产模式的记录与研究,描绘了在工业体系下新的生产关系如何重新塑造了新的城市空间关系,生产模式的批量化与个人化同时介入城市,实现了物体之间、空间之间、人力之间的重新连接。

"Interconnected in Production" is a research in view of production models in contemporary China. By observation and analysis of production patterns in the Yangtse Delta region, this section depicts how urban space is reshaped by new production relations where mass-production go hand in hand with customization and objects, spaces and spaces are interconnected.

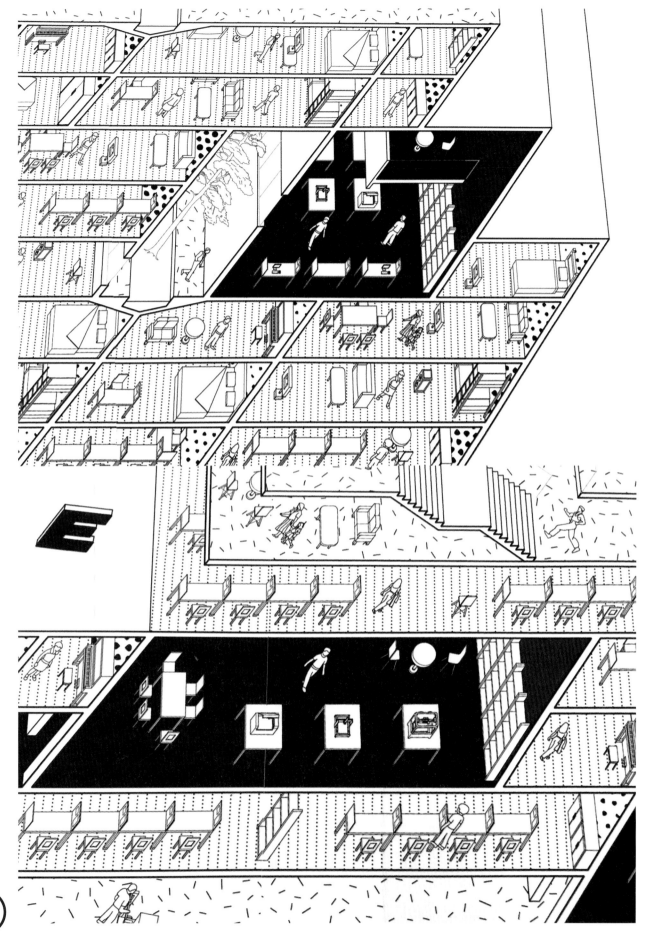

13

展项名称：体验 Hyperloop TT
Exhibition Title: Hyperloop TT Experience

策展人：李翔宁，邓圆也
Curators: LI Xiangning, DENG Yuanye

参展人：Hyperloop Transportation Technologies
Exhibitor: Hyperloop Transportation Technologies

我们致力于展现HyperloopTT的三个方面——移动性、连接性和可持续性。当今社会大多数交通方式已经过时,不仅过度负荷,并且需要大额的维护费用。太多的城市正经受着交通网络带来的空气污染。对于一个更加绿色、效率更高的交通系统的需求迫在眉睫。Hyperloop是一种利用管道连接城市并运送乘客和货物的交通系统。真空管道加上磁悬浮推进技术让管道中的客舱不受摩擦力,平稳达到与飞机媲美的速度。这些操作的动力均来自于绿色能源以及存储系统。我们可以通过建造Hyperloop创造更多的能源。我们的团队包括来自不同背景的资深工程师、科学家和创想家。我们发明并获得了独有的技术与科技专利,800多名员工遍布38个国家。我们已经与美国加州、斯洛伐克、阿布扎比酋长国、智利共和国、法国、印度尼西亚和韩国签约,让Hyperloop成为当今社会最安全的交通方式。交通的未来比你想象得离你更近。

We focus on the three keywords of Hperloop: mobility, sustainability and connectivity. Most current mass transport systems are outdated, overburdened and costly to maintain, while many cities are plagued by traffic congestion and poor air quality. The need for greener and more efficient transportation has never been greater. At its core, hyperloop™ is a tube-based inter and intra-city system for transporting passengers and cargo. Drastic reduction of air in the tube along with magnetic levitation and propulsion allows capsules to move through the tube with nearly zero friction and safely accelerate to airplane speeds. This is all powered by a combination of alternative energy and energy conservation systems. We are building hyperloop to produce as much or more energy than it uses. We are engineers, scientists and creative thinkers with various backgrounds and experiences, public or private. We have grown to over 800 professionals working across 38 countries. We have invented, patented and licensed technologies, signed agreements to build in California, Slovakia, Abu Dhabi, the Czech Republic, France, Indonesia and Korea. We've also been invited to consider several other contracts. We're working to make hyperloop the safest form of transportation today. The future of transportation is closer than you think.

展项名称：数字建造：数字金属
Exhibition Title: Digital Fabrication: Digital Metal

策展人：袁烽
Curator: Philip F. Yuan

参展人：马尼亚·阿格海伊·梅伯蒂，本杰明·迪伦伯格 / 苏黎世联邦理工学院建筑系数字建造技术中心
Exhibitors: Mania Aghaei Meibodi, Benjamin Dillenburger / ETH Zurich Digital Building Technologies

"数字金属"是一座在节点上采用非标准化设计的空间结构，所有节点均通过3D打印制模，再由金属铸造完成。3D打印的模具塑造了金属的最终形态，使定制复杂形态的结构构件成为可能，并且高度还原了表面的各种细节。在"数字金属"中有超过200个定制节点，在节点设计过程中采用了特殊的算法，这种几何上的复杂性贯彻于节点的设计与生成逻辑。超过400m长的铝管成品在简单加工之后就能安装，进一步提升建造的自动化程度。

"数字金属"是世界上第一座结合金属铸造和3D打印翻模的建筑结构，它向我们展示了数字工具及其对建筑、对真正建造案例的影响与应用，也打开了建筑形式自由的可能。

Digital Metal is a space frame structure with non-repetitive custom metal joints which are casted through 3D printed molds. Shaping metal through 3d printed molds allows the fabrication of bespoke structural building parts from metal with complex inner and outer features as well as surfaces with rich details. Digital Metal space frame consists of over 200 customized joints. Specific algorithms are developed to generate the frame and joints under structural and fabrication considerations. Over 400 meters of off-shelf aluminum profiles can cut to precise lengths with a simple robotic maneuver, helping to increase the level of automation in fabrication.

This is the first architectural structure made of hybrid cast metal and 3D printed molds in the world, offering a unique way of thinking about digital tools, their implications, applications and significance for real-world architecture and construction.

展项名称：数字建造：超薄纸板大跨建构
Exhibition Title: Digital Fabrication: Shells with Thin Sheet Materials

策展人：袁烽
Curator: Philip F. Yuan

参展人：王祥，一造科技
Exhibitors: WANG Xiang , Fab-Union

"超薄纸板大跨建构"是对轻型材料和大跨度结构的一次尝试。整个壳体结构的设计应用实验了一种新型的、针对超薄板材的壳体结构设计概念——细胞化墙体结构，尝试利用1mm的超薄纸板材料搭建大型的壳体结构，是一种基于材料性能化的设计方法，也是向砌筑这门古老技艺的重新回溯。

整个结构是由179个纸板单元组成的大跨空间。整个壳体结构通过结构找形技术得出整体的、仅受压的平衡系统。同时，通过引入腔体结构的局部膜系统，来提高结构局部的刚度，确保壳体的稳定性。

The shell with thin sheet material is a structural experiment of an innovative structure concept—the cellular cavity structure. The design and generation of the structure concept follows an advanced structure design method based on material performance. It enables a simple fabrication and assembly process of a large-span shell structure with super thin material.

The whole structure is built with 179 cellular elements, fabricated by only a 1mm paperboard. The geometry of the shell is defined by a state-of-the-art form-finding technique to maintain a global state of equilibrium and its local stability is enhanced by membranes of the cavity form.

展项名称：数字建造：机器人木构
Exhibition Title: Digital Fabrication: Robotic Timber Construction

策展人：袁烽
Curator: Philip F. Yuan

参展人：袁烽，柴华，孟刚，一造科技
Exhibitors: Philip F. Yuan, CHAI Hua, MENG Gang, Fab-Union

"机器人木构"项目旨在探索结构性能化建筑设计方法和机器人建造技术在木构建筑设计中的应用。项目关注结构原型与几何形式之间的关联性。以木网壳结构为原型,项目通过结构性能模拟与优化技术对网壳结构的结构尺寸进行了优化。结构性能的连续变化被转化为网壳结构的形式微差,从而在几何形式和结构性能之间建立了直接关联。以建筑机器人为工具,项目采用机器人线切割技术实现了网壳结构的数字建造。材料特性、结构性能、建筑几何与建造约束等因素被有机融合在空间木网壳结构的设计与建造过程中。

The "Robotic Timber Construction" project explores the Structural Performance-based Design Methods and Robotic Fabrication approaches in timber architecture. This project focuses on the correlation between structural performance and architectural geometries. Taking timber gridshells as structural prototypes, the gridshell sizes are optimized with structural simulation and optimization techniques. The gradients of structural performance are transformed into a continuous change in the form of a gridshell structure, resulting in a direct correlation between geometries and structural performance. In the fabrication stage, robotic wire cutting techniques are employed to ensure a precise presentation of the design. Material properties, structural performance, form-finding methods and fabrication constraints are integrated into the design and construction of the spatial gridshell structure.

今天，"上海都市范本"意味着什么？当城市发展方式从"外延式"转换为"内涵式"，城市更新方式也在经历一场从已有范式到新范式的转换过程——由上至下的更新模式正在被上下结合的新模式代替。由政府主导的浦江两岸贯通试图从城市整体的角度缝合原本断裂的城市空间；由建筑师与市民自发参与的微更新以小规模、低影响的渐进式改善方式，意图从更细微的断点入手缝补社区空间网络。城市空间的更新折射了市民生活方式和城市发展方式的变化，只要城市存在，城市更新就永远在过程中。

What is the logic of today's "Shanghai Sample"? When the urban development model changes from "denotative" to "connotative", city regeneration undergoes a transition from the existing paradigm to a new one - a model combining top-down and bottom-up methods, instead of the traditional "top-down". There comes the Waterfront Connection, promoted by the government, with attempts to sew up the long-segregated urban space from a macro view of the city and micro-regenerations springing up from architects and citizens to repair the network of community space. The micro approach involves small-scale, low-impact interventions that connect minor breakpoints. The renewal of spaces reflects changes in lifestyle and urban development. As the city continues to thrive, urban renewal will always be in the progress.

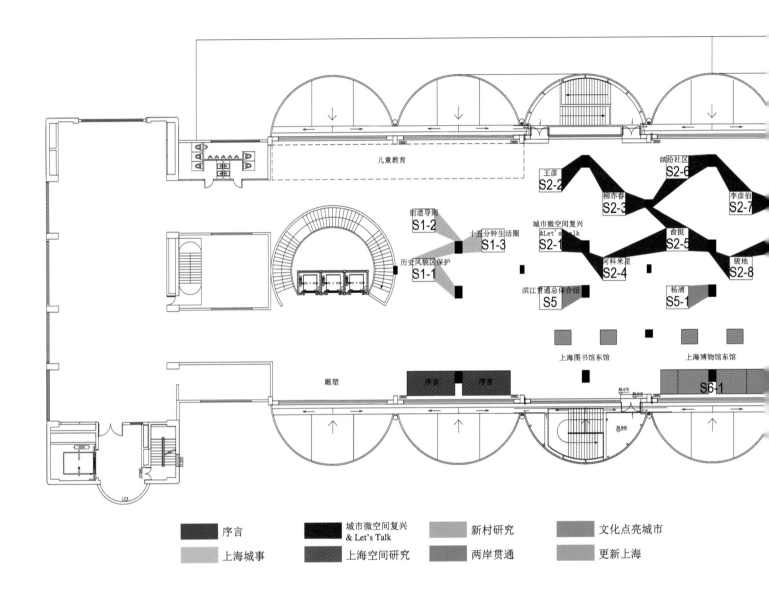

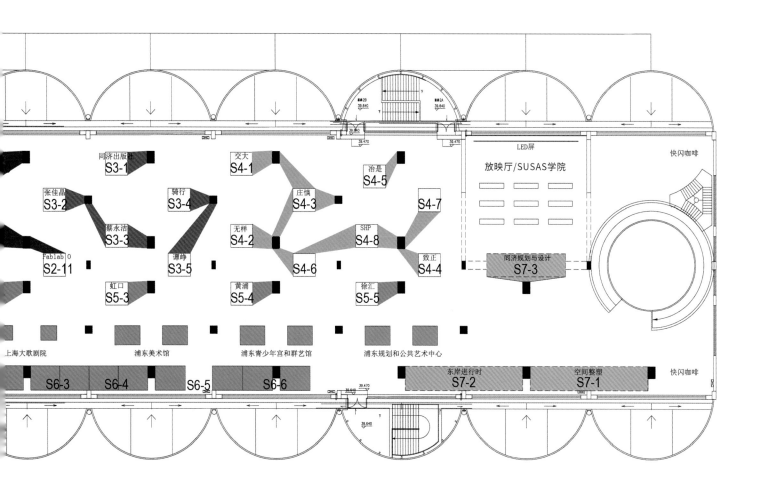

上海都市范本
策展人：支文军，戴春，冯路
策展团队：时代建筑、Let's Talk 学术论坛
策展助理：陈海霞，付一然，金怡，陈婷，曾婧如，周姝青，曹书韵，王佳佳（展陈设计）
特别鸣谢：上海同济规划设计研究院，上海风语筑展示股份有限公司，上海极臻三维设计有限公司，筑竹空间

Shanghai Sample, 2017SUSAS
Curators：ZHI Wenjun, DAI Chun, FENG Lu
Curatorial Team: *Time +Architecture*, Let's Talk Forum
Curatorial assistants: CHEN Haixia, FU Yiran, Crystal Jin, CHEN Ting, ZENG Jingyu, ZHOU Shuqing, CAO Shuyun, WANG Jiajia (Exhibiiton Design)
Acknowledgement: Shanghai Tongji Urban Planning & Design Institute, Shanghai Fengyuzhu Exhibition Co., Ltd.,
XUBERANCE DESIGN Co., Ltd. ,BAMBUSPACE

由政府发起的城市更新举措，自上而下调节城市空间的发展，进而创造和谐人居环境。大方向的调整与小规模的更新互相咬合形成紧密的齿轮——使上海城市的发展进程更加稳步向前推进。同时，对城市来说，城市更新本身就是一种"连接"——是一座连接城市的过去、现在与未来的桥梁。展览着眼于那些随着上海城市发展的喧嚣而逐渐沉默的区域，用当下的眼光关照场所历史中的生发与消亡，从而观察过去、实践当下，并进一步反思在城市更新、未来美好的愿景下不断涌现的、真实的建筑学与城市问题。

The top-down government-initiated urban renewal efforts are meant to adjust the development of urban spaces, thus creating a harmonious living environment. Sweeping transformations and micro interventions form a close-knitted set of gears, pushing the Shanghai's development process forward in steady paces. At the same time, urban regeneration itself is a kind of "connection", bridging the past, present and future. This exhibition focuses on decaying areas, following the birth and disappearance of history from contemporary perspectivesve, in order to reflect on the past, to keep exploring and practicsing, and to rethink the constantly emerging architecture and urban issues in the light of urban regeneration.

展项名称：上海城市历史风貌保护
Exhibition Title: Shanghai Historic Preservation

参展人：上海市规划和国土资源管理局
Exhibitor: Shanghai Municipal Bureau of Planning and Land Resources

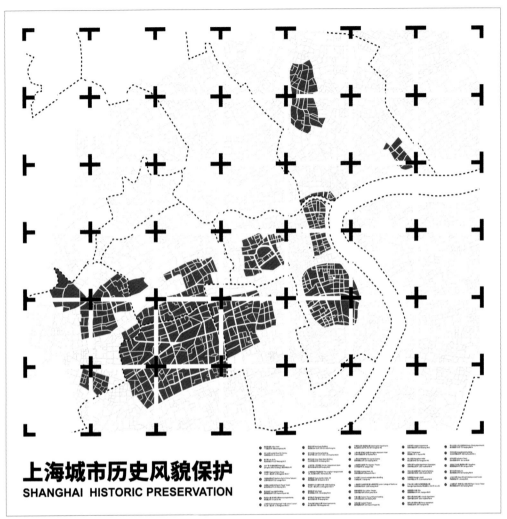

展项名称：上海市街道设计导则
Exhibition Title: Shanghai Street Design Guidelines

参展人：上海市规划和国土资源管理局
Exhibitor: Shanghai Municipal Bureau of Planning and Land Resources

安全街道 Safe street
车辆各行其道、有序交汇、安宁共享，保障各种交通参与者人身安全，保障交通活动有序进行。

绿色街道 Green street
促进土地资源集约、节约，倡导绿色低碳，鼓励绿色出行，增进居民健康，促进人工环境与自然环境和谐共存。

交通有序 Orderly traffic
协调人、车、路的时空关系，促进交通有序运行。

慢行优先 Non-motorized priority
维持街道的人性化尺度与速度，社区内部街道宁静共享。

步行有请 Walking space for pedestrians
为行人提供宽敞、畅通的步行通行空间。

过街安全 Safe crossings
提供直接、便利的过街可能，保障行人安全、舒适通过路口或横过街道。

骑行顺畅 Continuous bicycle lane
保障非机动车，特别是自行车行驶路权，形成连续、通畅的骑行网络。

设施可靠 Reliable facilities
提供可靠的街道环境，增加行人安全感。

资源集约 Resource efficiency
集约、节约、复合利用土地与空间资源，提升利用效率与效益。

绿色出行 Green transport
倡导绿色出行，鼓励步行、自行车与公共交通出行。

生态种植 Ecological planting
提升街道绿化品质，兼顾活动与景观需求，突出生态效益。

绿色技术 Green technology
对雨水径流进行控制，降低环境冲击，提升自然包容度。

活力街道 Vibrant street
提供开放、舒适、易达的空间环境体验，增进市民交往交流，提升社区生活体验，鼓励创意与创新。

智慧街道 Smart street
整合街道设施进行智能改造，提升智行协助、安全维护、生活便捷、环境"智"理服务。

功能复合 Mixed-use functions
增强沿街功能复合，形成活跃的空间界面。

活动舒适 Comfort activity zone
街道环境舒适、设施便利，适应各类活动需求。

空间宜人 Pleasant space
街道空间有序、舒适、宜人。

视觉丰富 Rich vision experience
沿街建筑设计应满足人的视角和步行速度视觉体验需求。

风貌塑造 Street characteristic
街道空间环境设计注重形成特色，塑造地区特征，展现时代风貌。

历史传承 History inheritance
依托街道传承城市物质空间环境，延续历史特色与人文氛围。

设施整合 Facility integration
智能集约化改造街道空间，智慧整合更新街道设施。

出行辅助 Transport aid
普及智能公交、智能慢行，促进智慧出行，协调停车供需。

智能监控 Smart monitoring
实现监控设施全覆盖、呼救设施定点化，提高安全信息传播的有效性。

交互便利 Convenient information interaction
设置信息交互系统，促进社区智慧转型。

环境智理 Smart environmental stewardship
加强环境检测保护，促进智能感应并降低能耗。

展项名称：上海15分钟社区生活圈
Exhibition Title: Shanghai Planning Guidance of 15-minute Community-life Circle

参展人：上海市规划和国土资源管理局
Exhibitor: Shanghai Municipal Bureau of Planning and Land Resources

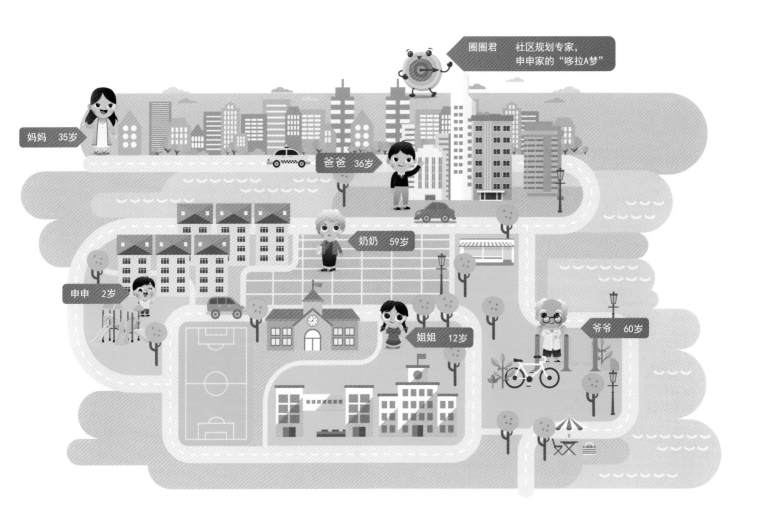

2010年至今，上海已经走入稳步发展的渐进式更新阶段。在这里，政府与社区的协同合作开始发挥更大作用，和谐社会网络的构建也成为今天城市更新中最重要的一环。除了自上而下的宏观调整，自下而上的"微更新"也是不可或缺的一部分。城市微更新以适应新的日常生活与工作的需求为导向，对一系列片段化的城市建成环境和既有建筑进行调整型更新，是更贴近空间使用者的更新行为。建筑师与市民自发参与，以小规模、低影响的渐进式改善方式缝补社区空间网络。在存量建成环境语境下，微更新更是对城市中"失落的空间"的再思考与再创造，2015年，Let's talk学术论坛创办人俞挺和戴春发起了"城市微空间复兴计划"，号召设计师研究身边的空间并开展微更新实践，这里展出实践中的部分案例，展现了一种对城市更新方式的全新认知。

Since 2010, Shanghai has entered into the stage of "gradual regeneration". At this point, the collaboration of government and community begins to play a greater role. The construction of a harmonious social network is also becoming the most important part of urban regeneration. In addition to the top-down macro adjustment, bottom-up "micro-regeneration" is another integral part. Urban micro-regeneration aims to satisfy modern demands for life and work. It is a kind of regenerative behaviour that focuses on adjusting series of fragmented urban built environments to better serve their users. Architects and the public are spontaneously participating in small-scale, low-impact, progressive improvements to sew up the community space network. In the context of the stock, or completed environment, micro-regeneration means to rethink and re-create the city's "lost spaces". In 2015, Yu Ting & Dai Chun, co-founders of the Let's talk forum, launched the "Urban humble-space Regeneration Plan", encouraging designers to examine their neighborhood and to carry out micro-regeneration. The exhibition shows some examples of their practices, presenting a new understanding of the methods of urban regeneration.

展项名称：城市微空间复兴计划 & Let's talk
Exhibition Title: Urban Humble-Space Regeneration & Let's talk
参展人：Let's talk 学术论坛
Exhibitor: Let's talk Academic Forum

展项名称：城市泡泡 / 韧山水 / 大烟囱咖啡馆
Exhibition Title: City Bubble / Flexible Landscape / Chimney cafe
参展人：王彦
Exhibitor: WANG Yan

展项名称：例园茶室
Exhibition Title: Tea House in Li Garden
参展人：大舍建筑设计事务所
Exhibitor: Atelier Deshaus

展项名称：社区活力发生器 / 刷新城中村
Exhibition Title: Community Vitality Regenerator / Refreshing Urban Village
参展人：李彦伯 / 同济大学普方研究室
Exhibitors: LI Yanbo / Tongji University, Profound Lab

展项名称：社区，让生活更美好——浦东缤纷社区
Exhibition Title: Site Project of Pudong Colorful Community
参展人：上海市城市规划设计研究院
Exhibitor: Shanghai Urban Planning & Design Research Institute

展项名称：八分园及 Wutopia Lab 的城市印迹
Exhibition Title: Eight Tenths Garden & City Marks of Wutopia Lab
参展人：俞挺工作室
Exhibitor: Wutopia Lab

展项名称：新天地临时读书空间
Exhibition Title: Temporary Reading Pavilion
参展人：上海阿科米星建筑设计事务所有限公司
Exhibitor: Atelier Archmixing

展项名称：重"石"乡愁——城市，走向新社区
Exhibition Title: Shiquan Road Street Urban Renewal
参展人：上海骏地建筑设计咨询股份有限公司
Exhibitor: JWDA (Joseph Wong Design Associates)

展项名称：旧里新厅
Exhibition Title: SL Space-Micro Regeneration in West Guizhou Lilong
参展人：童明工作室（上海梓耘斋建筑设计咨询有限公司）
Exhibitor: TM Studio

展项名称：方寸漫游
Exhibition Title: Roaming Streets and Lanes
参展人：席闻雷
Exhibitor: XI Wenlei

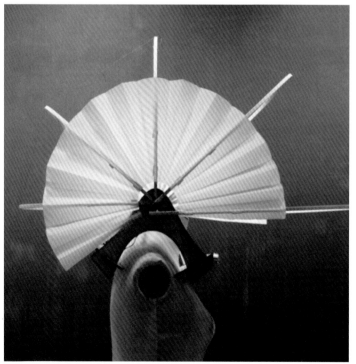

展项名称：机器制造机器"及 Fablab O 中国"数制"城市智造 2.0
Exhibition Title: "Machine-Makes-Machines" and Fablab O Urban Intelligent Manufacture 2.0
参展人：中国"数制"工坊
Exhibitor: Fablab O

S3

展项名称：上海空间研究
Exhibition Title: Shanghai Urban-Space Research

策展人：戴春，冯路
Curators: DAI Chun, FENG Lu

参展人：王卓尔，蔡永洁，张佳晶，谭峥，同济大学出版社
Exhibitors: WANG Zhuoer, CAI Yongjie, ZHANG Jiajing, TAN Zheng, Tongji University Press

上海空间研究展览集合了5项从不同角度出发对城市空间的再思考与新实验。"骑行上海"板块通过对公共空间再分配及构想，意图为未来的上海提供更好的骑行环境。

虹口滨江基础设施更新计划选择了虹口滨江的四处典型地段，以景观都市主义的"地形策划"理念为基本方法，提出了对基础设施所构成的公共空间进行干预更新的策略。

城市填空 & 城市缝合试图系统性改善市政存量，使其成为增量，而"缝合"意为缝合被"撕裂"的城市，改善早期城市规划里出现的"大马路中心城"模式。

陆家嘴改造则通过日常空间的塑造演示了一次新城再城市化的实验，在保留陆家嘴宏大景象的同时，试图营造出崭新的市民化的城市体验。这些探索与实验拓展了城市空间发展方式的可能性，构建了一幅充满活力的城市未来意象。

Shanghai Urban-Space Research Includes five projects that rethinks and experiments on urban spaces from their own angles. Cycling@Shanghai intends to provide a better riding environment for Shanghai by redistributing and re-planning the city's public spaces.

The hongkou District waterfront infrastructure renewal plan chooses four typical sections along the hongkou waterfront for infrastructure regeneration initiatives. Applying the concept of "landscape planning" of Landscape Urbanism, regeneration strategies on the public spaces are proposed, especially infrastructure.

Urban Infill & Urban Seaming attempts to systematically improve the completed urban spaces and give them the potentials to expand. "Infill" refers to improve the density of the city center, to promote urbanization by properly filling it with functional elements. "Seaming" refers to sew up a torn city, to improve on the "mega-road-centered city" model common among earlier city plans.

The Lujiazui Regeneration demonstrates the experiment of reurbanization through shaping common spaces, and tries to create a new civic city experience while maintaining the grand narration of Lujiazui. These explorations and experiments have expanded the possibility of urban spatial development and helped create a new vibrant image of the city's future.

展项名称：带你一起飞
Exhibition Title: Up With the Books
参展人：同济大学出版社
Exhibitor: Tongji University Press (TJUP)

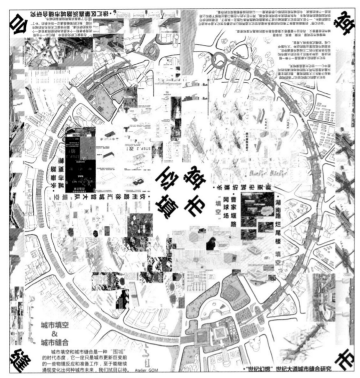

展项名称：城市填空 & 城市缝合
Exhibition Title: Urban Infill & Urban Seaming
参展人：张佳晶 / 上海高目建筑设计咨询有限公司
Exhibitors: ZHANG Jiajing / Atelier GOM

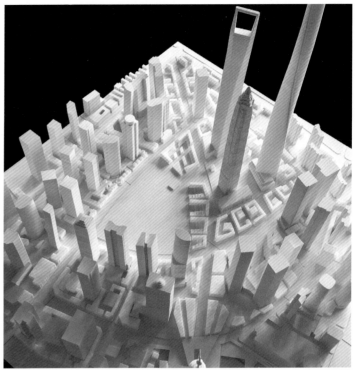

展项名称：新城再城市化：陆家嘴实验
Exhibition Title: Reurbanization of Newtown: Experiment of Lujiazui
参展人：蔡永洁（同济大学教授）同济大学建筑系 2017 本科毕业设计"新城再城市化：陆家嘴实验"课题组 同济大学 / 乔治亚理工中美联合生态化城市设计实验室
Exhibitors: Cai Yongjie (Professor of Tongji University)
Working team of 2017 undergraduate thesis project "Re-urbanization of New Town: Experiment in Lujiazui" architectural faculty of Tongji University, SINO-U.S. ECO URBAN LAB - Tongji University & Georgia Institute of Technology

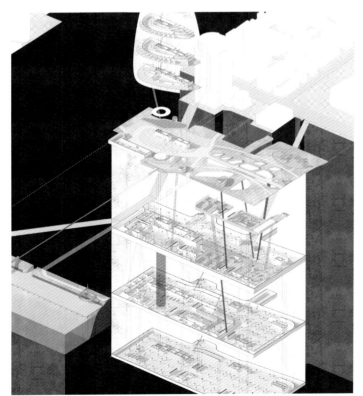

展项名称：虹口滨江基础设施更新计划
Exhibition Title: Planning the Infrastructural Renewal for the Hongkou Waterfront
参展人：谭峥
Exhibitor: TAN Zheng

展项名称：骑行上海 I：从路权至道路更新
Exhibition Title: Cycling@Shanghai I: A Renovation from Right of Way
参展人：王卓尔
Exhibitor: WANG Zhuoer

S4

展项名称：新村研究
Exhibition Title: Study of New Villages

策展人：冯路
Curator: FENG Lu

参展人：无样建筑工作室，致正建筑工作室，阿科米星建筑设计事务所，上海交通大学建筑系，冶是建筑工作室
Exhibitors: Wuyang Architecture, Atelier Z+, Atelier Archmixing, Department of Architecture of Shanghai Jiaotong University, Atelier YeAS

新村研究项目包括"上海计划"研究系列和冶是建筑工作室的新泾诸村案例研究。其中，上海计划（简称 SHP）研究项目开始于 2014 年上半年，是由冯路、张斌、庄慎、范文兵四位建筑师共同发起的一项长期计划，意图通过对上海日常城市空间的再研究而进一步思考、发现和认识我们在当代所面临的建筑学和城市问题。

工人新村是上海城市空间的重要内容。1949 年建国后上海转变成为重要的工业城市，为大量的工人解决居住问题成为当时城市发展的重要使命。建造于 20 世纪 50—70 年代的工人新村是社会集体主义生产和组织方式的空间化。伴随着 90 年代后上海经济和社会的转变，原有的集体单位逐步解体或者迁移，工人新村经历了由集体空间到共有空间乃至城市社区的自我转变。对工人新村空间状况的调查和图解因此成为一项有意义的工作，对旧城的有机更新可以提供有价值的认识、思考和文献。

The cases of Study of New Villages include the Shanghai Project (ShP) series and Xinjing New Villages research conducted by Atelier YeAS. Shanghai Project is a research project that began in early 2014, which was a long-term plan co-sponsored by four architects: Feng Lu, Zhang Bin, Zhuang Shen and Fan Wenbing. Their intention was to discover and understand contemporary architectural and urban issues by researching and rethinking commonplace urban spaces in Shanghai. Xinjing New Villages is studied by Atelier YeAS.

Workers' New Village is a significant content of Shanghai urban spaces. After the founding of the People's Republic of China in 1949, Shanghai became an important industrial city when the problem of accommodating large groups of workers became an important task of urban development. Workers' New Villages, built in the 1950s-1970s, used to be the spatial embodiment of collectivism prevailing in production and social organization. As Shanghai went through a major economic and social transformation in the 1990s, the original collective units gradually moved or disintegrated. Meanwhile, Workers' New Villages was transferred from a collective unit to a common housing area. The visualized investigation of the spatial condition of the Workers' New Villages is a meaningful project for it provides records, knowledge and insights about the organic regeneration of old city.

展项名称：上海计划 / 新村研究爆炸图
Exhibition Title: Shanghai Project/ Explosion of the New Village
参展人：庄慎，唐煜，陈平楠，黄泽填，王子潇，阿科米星建筑事务所
Exhibitors: ZHUANG Shen, TANG Yu, CHEN Pingnan, HUANG Zetian, WANG Zixiao, Atelier Archmixing

展项名称：上海计划 / 航运新村研究
Exhibition Title: Shanghai Project / a Study of Hangyun New Village
参展人：冯路，郑思宇，周铭迪，李璐，无样建筑工作室
Exhibitors: FENG Lu, ZHENG Siyu, ZHOU Mingdi, Li Lu, Wuyang Architecture

展项名称：上海计划 / 定海路 449 弄产业工人社区研究
Exhibition Title: Shanghai Project/ a Study of Dinghairoad 449 Lane, an Industrial Urban Workers Residential Community
参展人：范文兵，张帆，安康，张雨薇，上海交通大学建筑学系 + 思作设计工作室
Exhibitors: FAN Wenbing, ZHANG Fan, AN Kang, ZHANG Yuwei, Department of Architecture, Shanghai Jiao Tong University + ATELIER FAN

展项名称: 上海计划／田林新村研究
Exhibition Title: Shanghai Project / a Study of Tianlin New Village
参展人: 致正建筑工作室／张斌，张雅楠，孙嘉秋，徐杨，许晔，肖燕萍，于瑞莹，谢兆荣
Exhibitors: Atelier Z+ / ZHANG Bin, ZHANG Yanan, SUN Jiaqiu, XU Yang, XU Ye, XIAO Yanping, YU Ruiying, XIE Zhaorong

展项名称: 上海计划／新泾诸村研究
Exhibition Title: A Study of New Villages Beixinjing
参展人: 冶是建筑工作室／周渐佳，李丹峰，叶之凡，孙正宇，唐奕诚，聂方达
Exhibitors: YeAS Studio / ZHOU Jianjia, LI Danfeng, YE Zhifan, SUN Zhengyu, TANG Yicheng, NIE Fangda

在物质空间的更新上，我们同时扮演着使用者与创造者的角色：本能地去创造适应人类使用需求的空间。黄浦江作为上海人民的母亲河，生生不息地记录着城市前进的足迹。在过去，上海滨水空间存在着许多"断点"，为了营造更加连续开放、可达舒适的高品质公共空间，两岸贯通的新举措试图"连接"这些断点，同时塑造优美亮丽的景观形象，展现深厚多样的历史风貌，打造一条亲水宜人的、服务于全体市民的绿色生活岸线，并真正提升黄浦江两岸开放空间的潜在价值。相信通过城市生活与滨江空间交织互动，我们的当下将与未来更好的城市生活空间紧密联系在一起。

In regenerating physical spaces, we simultaneously play the roles of the user and the creator: instinctively creating space to meet the needs of mankind. Huangpu River, as the mother river of Shanghai, endlessly documents the city's footsteps. In the past, there were many breakpoints along the waterfront of Shanghai. To make more open, reachable and comfortable public spaces, new actions are carried out to connect these breakpoints: designing beautiful landscape and presenting historical features; creating a waterfront, pleasant, green shoreline for all citizens, meanwhile enhancing potential value on both banks of the Huangpu River. The intertwined interaction of urban life and riverside space is sure to make better urban spaces in the future.

展项名称：浦东新区滨江贯通
Exhibition Title: Pudong Waterfront Connection
参展人：浦东新区
Exhibitor: Pudong New Area

展项名称：杨浦区滨江贯通
Exhibition Title: Yangpu Waterfront Connection
参展人：杨浦区
Exhibitor: Yangpu District

展项名称：虹口区滨江贯通
Exhibition Title: Hongkou Waterfront Connection
参展人：虹口区
Exhibitor: Hongkou District

扬子江码头段（500米）　　国客中心段（800米）　　置阳段（400米）　　国航中心段（800米）

展项名称：黄浦区滨江贯通
Exhibition Title: Huangpu Waterfront Connection
参展人：黄浦区
Exhibitor: Huangpu District

展项名称：徐汇区滨江贯通
Exhibition Title: Xuhui Waterfront Connection
参展人：徐汇区
Exhibitor: Xuhui District

S6

展项名称：文化点亮城市 / 策展人：支文军，戴春
Exhibition Title: Culture Englightening the City / Curators: ZHI Wenjun, DAI Chun

参展人：浦东新区规划和土地管理局，上海市文化广播影视管理局，丹麦SHL建筑事务所，大卫·奇普菲尔德建筑事务所，同济大学建筑设计研究院（集团）有限公司—若本建筑工作室，中国建筑设计研究院有限公司—李兴钢建筑工作室，斯诺赫塔，奥唐纳和托米建筑事务所，让·努维尔事务所，山水秀建筑事务所，同济大学建筑设计研究院（集团）有限公司—原作设计工作室，MVRDV
Exhibitors: Pudong New Area Planning and Land Resource; Shanghai Municipal Administration of Culture, Radio, Film and TV; Schmidt Hammer Lassen Architects; David Chipperfield Architects; Rurban Studio, Tongji Architectural Design (Group) Co., Ltd.; Atelier Li Xinggang, China Architecture Design & Research Group Co., Ltd.; Snøhetta; O'Donnell + Tuomey Architects; Ateliers Jean Nouvel; Scenic Architecture Office; Original Design, Tongji Architectural Design (Group) Co., Ltd.; MVRDV

当今社会需要释放城市的人性化温度，从注重经济价值到注重文化价值，这个转变正回应了公众内心真实的深层诉求。城市不仅是物质空间的集合，也是人类的精神家园。以"文化点亮城市"为主题——浦东三大市级文化设施与四个区级文化设施的建设意图缝合城市文化网络，使两岸的文化场所能同时多方面、多角度地服务于上海市民的精神生活。新的文化建筑在创造更集中、受众更广泛的城市生活共享平台的同时，向世界展示了上海不仅是不断飞跃的经济之城，也是独具魅力的文化之都。

Society demands more humane ideals in cities today. Switching emphasis from economic value to cultural value, the government is responding to the demands of the public. The city is definitely not only a collection of physical space, but also the spiritual paradise of mankind. Themed as "Culture Enlightens the City", the construction of cultural facilities in Pudong aims to improve the cultural network of the city, so that cultural sites on both banks can simultaneously provide multi-faceted services to fulfill the spiritual needs of the residents in Shanghai. New cultural buildings provide the public with a shared platform more concentrated and inclusive. At the same time, it shows to the whole world the image of Shanghai as a global city with charming culture, in addition to its impressive economic development.

展项名称：上海图书馆东馆
Exhibition Title: Shanghai Library (East Hall)
参展人：丹麦 SHL 建筑事务所，大卫·奇普菲尔德建筑事务所
Exhibitors: Schmidt Hammer Lassen Architects, David Chipperfield Architects

展项名称：上海博物馆东馆
Exhibition Title: Shanghai Museum (East Hall)
参展人：同济大学建筑设计研究院（集团）有限公司–若本建筑工作室，中国建筑设计研究院有限公司–李兴钢建筑工作室
Exhibitors: Rurban Studio, Tongji Architectural Design (Group) Co., Ltd.;
Atelier Li Xinggang, China Architecture Design & Research Group Co., Ltd.

展项名称：上海大歌剧院
Exhibition Title: Shanghai Grand Opera House
参展人：斯诺赫塔，奥唐纳和托米建筑事务所
Exhibitors: Snøhetta, O'Donnell + Tuomey Architects

展项名称：浦东美术馆
Exhibition Title: Pudong Museum of Art
参展人：让·努维尔事务所，大卫·奇普菲尔德建筑事务所
Exhibitors: Ateliers Jean Nouvel, David Chipperfield Architects

展项名称：浦东新区青少年活动中心及群众艺术馆
Exhibition Title: Pudong New District Youth & Children's Center and Mass Art Center
参展人：山水秀建筑事务所，同济大学建筑设计研究院（集团）有限公司–原作设计工作室
Exhibitors: Scenic Architecture Office; Original Design, Tongji Architectural Design (Group) Co., Ltd.

展项名称：浦东新区青少年活动中心及群众艺术馆
Pudong New District Youth & Children's Center and Mass Art Center

展项名称：浦东规划和公共艺术中心
Exhibition Title: Pudong Urban Planning and Public Art Center
参展人：大卫·奇普菲尔德建筑事务所，MVRDV
Exhibitors: David Chipperfield Architects, MVRDV

S7

展项名称：更新上海
Exhibition Title: Renewing Shanghai

策展人：戴春
Curator: DAI Chun

参展人：上海东岸投资（集团）有限公司，上海地产（集团）有限公司，同济规划与设计
Exhibitors: Shanghai East Bank Investment (Group) Co., Ltd., Shanghai Land (Group) Co., Ltd, Tongji Planning and Design

城市更新作为上海城市发展的关键词，已经成为上海城市发展创新转型的核心。上海城市要满足人们日益增长的生活需求、实现宜居城市和持续繁荣的发展目标，只有靠在现有空间内进行有效更新。在自上而下的城市更新中，国有建设企业发挥了重要作用，无论是地产集团围绕上海市委、市政府工作大局开展的系列更新实践，还是东岸集团基于浦东发展开展的黄浦江东岸沿线的建设，以及同济规划与设计机构的研究实践，都呈现了国有企业在上海城市更新中的创新与探索。

Urban regeneration, a keyword for urban development in Shanghai, has become the core of the city's development, innovation and transformation. To meet the citizens' growing need of life, to create a livable, sustainable and prosperous city, Shanghai is bound to implement effective regenerations on existing spaces. In the top-down urban regeneration process, the state-owned construction enterprises have played an important role: Shanghai Land Group has carried out a series of renewal practices responding to the overall plan made by the Shanghai municipal government; and Shanghai East Bund Investment group has conducted constructions along the east bank of the Huangpu River based on the city planning focus on Pudong Area; and the planning and design institutes] of Tongji University have focused their research and practices right in the city. All these works demonstrate the innovation and exploration efforts made in this urban regeneration progressbystate-owne denterprises.

展项名称：空间整塑
Exhibition Title: Space Reform
参展人：上海地产（集团）有限公司
Exhibitor: Shanghai Land (Group) Co., Ltd.

展项名称：东岸进行时
Exhibition Title: East Bund-ing
参展人：上海东岸投资(集团)有限公司
Exhibitor: Shanghai East Bund Investment (Group) Co., Ltd.

展项名称：有温度的城市——上海城市更新的同济实践
Exhibition Title: A City with Humanity-Tongji Practice of Urban Regeneration in Shanghai
参展人：同济大学建筑与城市规划学院 / 上海同济城市规划设计研究院 / 同济大学建筑设计研究院（集团）有限公司
Exhibitors: College of Architecture and Urban Planning, Tongji University, Shanghai Tongji Urban Planning & Design Institute, Tongji Architectural Design (Group) Co., Ltd

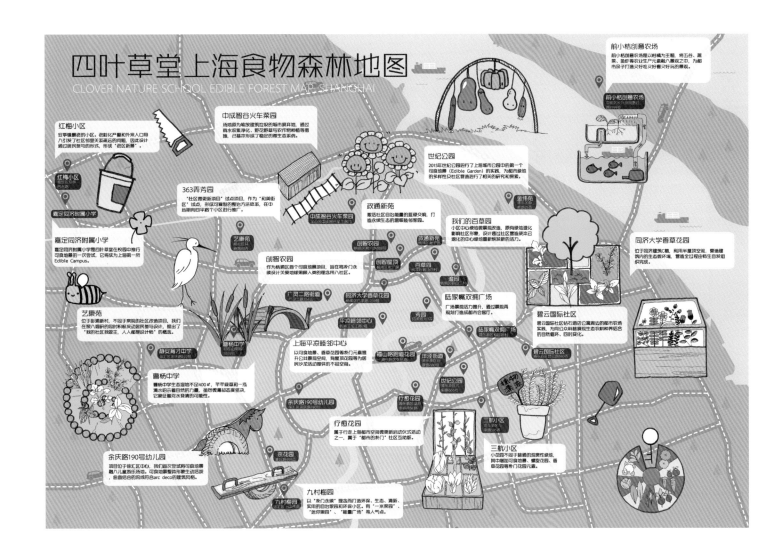

展览项目索引

	展项名称	策展人	参展人
070	P1/ 建造当代的文化图景	李翔宁 / 高长军	非常建筑等 27 个建筑设计事务所
076	P2/ 巴塞罗那：城市群大都会	胡安·布斯克茨，杨丁亮，尤拉利亚·戈麦斯·艾斯柯达	胡安·布斯克茨，杨丁亮，尤拉利亚·戈麦斯·艾斯柯达 / 巴塞罗那大都会区，哈佛大学设计学院
080	P3/ 拉斯维加斯工作室：来自罗伯特·文丘里和丹妮丝·斯科特·布朗档案馆的影像集	希拉·斯塔德勒 / 马提诺·斯泰尔利	比尔帕克博物馆
084	P4/ 液态历史：泰晤士河的想象与现实	大卫·尚贝尔，杰里米·蒂尔，马丁·维尔	中央圣马丁艺术与设计学院
088	P5/ 当代中国的多元建筑实践	李翔宁，高长军	王澍等 33 位建筑师
096	P6/ 与水共生：世界优秀水岸空间案例展	李翔宁，莫万莉，张子岳，邓圆也	BAUM 建筑师事务所，B.I.G，居依·诺丁森结构设计有限公司，JCFO，SASAKI，威尼斯建筑大学，维斯 / 曼弗雷迪建筑·城市·景观·设计事务所，West 8
106	P7/ 社会图景：来自城市内部的影像学	金江波	敖国兴，戴建勇，冯梦波，何崇岳，金江波，吉姆·斯皮尔斯，李消非，李振宇，马良，倪卫华，Nancy Royal，渠岩，沈少民，邵文欢，王川，王庆松，徐坦，杨泳梁，赵宏利，曾力
110	P8/ 木构与智构	方振宁	方振宁，林凡榆
114	P9/ 漫步环翠堂园景	方振宁	方振宁
118	P10/ 马列维奇视觉年表	方振宁	方振宁
124	P11/ 万象	方振宁	赵弘君，何汶玦，靳烈，王欣，马岩松，王昀，孟禄丁，李迪，李磊，方振宁，陈文令，吕越，张朝晖，FANGmedia
128	P12/ 鼓浪屿历史国际社区——共享遗产保护之路	方振宁	厦门市鼓浪屿——万石山风景名胜区管理委员会，北京国文琰文化遗产保护中心有限公司，清华大学建筑学院国家遗产中心
134	T1/ 林中之境	斯坦法诺·博埃里	斯坦法诺·博埃里
142	T2/2017 年欧盟当代建筑奖——密斯·凡·德·罗奖和 2016 年 "Fear of Column" 竞赛展览	李翔宁，高长军	密斯·凡·德·罗基金会
148	T3/ 全球建筑实践罗盘——一种新兴建筑的分类学	亚历杭德罗·扎拉-波罗，吉尔莫·费尔南德兹-阿巴斯卡尔	亚历杭德罗·扎拉-波罗，吉尔莫·费尔南德兹-阿巴斯卡尔
150	T4/ 哥伦布之谜系列雕塑展	戈勃朗基金会	戈勃朗基金会
152	T5/ 凝聚	方振宁	方振宁
156	T6/ 混乱中迷失	方振宁	劳伦斯·维纳
162	T7/ 风卷	方振宁	王迈
164	T8/ 中国文人写意雕塑园（9 件）	方振宁	吴为山
170	T9/ 内省腔	郭晓彦	尹秀珍
174	T10/ 设计长椅	方振宁	靳烈
176	T11/ 风律	李翔宁	盛姗姗
178	T12/ 仓声·品	李翔宁	苏丹，王宁，张荐
180	T13/2340 洞	李翔宁	于幸泽

	展项名称	策展人	参展人
184	H1/ 回音：建筑与社会	冈萨雷斯，田唯佳	弗朗西斯·克雷, 荷西·拉艾多, 安德烈斯·哈克, 王子耕, 生态系统城市研究室, 乡村城市共建工作室, 白德拉研究所
190	H2/ 南京长江大桥记忆计划	鲁安东	LanD Studio
194	H3/ 濑户内国际艺术节	北川富朗	林舜龙
198	H4/ 连接：空间移动	郭晓彦	刘韡, 沈远, 邱志杰, 徐震, 黄永砯, 何岸, 王郁洋, 奥拉维尔·埃利亚松, 刘建华, 程然, 汉斯·欧普·德·贝克, 多米尼克·冈萨雷斯·福斯特, 法比安·纪劳德
214	H5/ 空间的边界	方振宁	王昀
218	H5/ 你是如此温柔、幻想速度	方振宁	黄莺
222	H5/ 寻找马列维奇	方振宁	张永和
226	H5/ 天路	方振宁	李磊
230	H5/ 蝴蝶夫人	方振宁	吕越
234	H5/ 小库——人工智能时代的未来都市	方振宁	何宛余 / 小库科技
238	H6/ 内心（影像）	方振宁	沈伟，马岩松
246	I1/ 桥	李翔宁，姚微微，谢雨晴	大舍建筑事务所, 无止桥慈善基金, 非常建筑事务所, 赫斯维克事务所, 詹姆士·康纳, 场域运作事务所, 马德里欧洲大学
258	I2/ 物联生产	一造科技	一造科技团队 / 冶是建筑工作室
262	I3/ 体验 HyperloopTT	李翔宁，邓圆也	Hyperloop Transportation Technologies
266	I4/ 数字建造：数字金属	袁烽	马尼亚·阿格海伊·梅伯蒂, 本杰明·迪伦伯格
268	I4/ 数字建造：超薄纸板大跨建构	袁烽	王祥 / 一造科技
270	I4/ 数字建造：机器人木构	袁烽	袁烽，柴华，孟刚 / 一造科技
276	S1/ 上海城事	支文军，戴春	上海市规划与国土资源管理局 / 上海同济规划设计研究院
281	S2/ 城市微空间复兴计划	戴春、俞挺	Let's talk 团队, 王彦, 柳亦春, 上海市城市规划设计研究院, 俞挺, 李彦伯, 阿科米星, 童明, 骏地, 席子, Fablab O
286	S3/ 上海空间研究	戴春，冯路	王卓尔, 蔡永洁, 张佳晶, 谭峥 / 同济大学出版社
289	S4/ 新村研究	冯路	无样建筑工作室, 致正建筑工作室, 阿科米星建筑设计事务所, 上海交通大学建筑系, 冶是建筑工作室
293	S5/ 两岸贯通	戴春	上海市规划与国土资源管理局 / 杨浦区, 浦东新区, 虹口区, 黄浦区, 徐汇区
300	S6/ 文化点亮城市	支文军，戴春	浦东新区规划和土地管理局, 上海市文化广播影视管理局, 丹麦 SHL 建筑事务所, 大卫·奇普菲尔德建筑事务所, 同济大学建筑设计研究院（集团）有限公司—若本建筑工作室, 中国建筑设计研究院有限公司—李兴钢建筑工作室, 斯诺赫塔, 奥唐纳和托米建筑事务所, 让·努维尔事务所, 山水秀建筑事务所, 同济大学建筑设计研究院（集团）有限公司—原作设计工作室, MVRDV
308	S7/ 更新上海	戴春	上海东岸投资（集团）有限公司, 上海地产（集团）有限公司

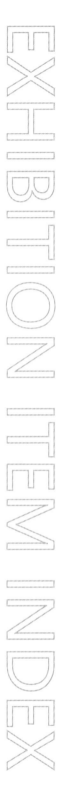

EXHIBITION ITEM INDEX

	Exhibition Title	Curator	Exhibitors
070	P1/ Constructing a Contemporary Cultural Landscape	LI Xiangning, GAO Changjun	FCJZ and other 26 ateliers
076	P2/ Barcelona, Metropolis of Cities	Joan·Busquets, YANG Dingliang, Eulalia·Gomez·Escoda	Joan Busquets, YANG Dingliang, Eulalia Gomez Escoda/ÀreaMetropolitana de Barcelona (AMB), Harvard Graduate School of Design (GSD)
080	P3/ Las Vegas Studio: Images from the Archives of Robert Venturi and Denise Scott Brown	Hilar Stadler, Martino Stierli	Museum im Bellpark
084	P4/ LIQUID HISTORIES:The Thames between imaginary and reality	David Chambers, Jeremy Till, Martyn Ware	Central Saint Martins with Illustrious
088	P5/ Diverse Practices in Contemporary Chinese Architecture	LI Xiangning, GAO Changjun	WANG Shu and other 33 architects
096	P6/ Living with Water: World Extraordinary Waterfront Space	LI Xiangning, MO Wanli, ZHANG Ziyue, DENG Yuanye	BAUM Architects, Bjarke Ingels Group, Guy Nordenson and Associates, James Corner Field Operations, SASAKI, Universita IUAV di Venizia, WEISS/MANFREDI Architecture/Landscape/Urbanism , West 8
106	P7/ Cases Social View: Iconography from City	JJN Jiangbo	AO GuoXing, DAI JianYong, FENG Mengbo, HE Chongyue, JJN Jiangbo, Jim Speers, LI Xiaofei, LI Zhenyu, MA Liang, NI Weihua, Nancy Royal, QU Yan, SHEN Shaomin, SHAO Wenhuan, WANG Chuan, WANG Qingsong, XU Tan, YANG Yongliang, ZHAO Hongli, ZENG Li.
110	P8/ Wooden Structures and Smart Structures	FANG Zhenning	FANG Zhenning, LIN Fanyu
114	P9/ Stroll in HuanCuiTang	FANG Zhenning	FANG Zhenning
118	P10/ Malevich Visual Chronology	FANG Zhenning	FANG Zhenning
124	P11/ Social View: Wanxiang	FANG Zhenning	ZHAO Hongjun, HE Wenjue, JIN Lie, WANG Xin, MA Yansong, WANG Yun, MENG Luding, LI Di, LI Lei, TONG Zhengang, FANG Zhenning, CHEN Wenling, Lyu Yue, ZHANG Zhaohui/FANGmedia.
128	P12/ Social View: A Sharing Conservation Approach:Gulangyu,a Historic International Settlement	FANG Zhenning	Xiamen Kulangsu Scenic Area Administrative Committee, Cultural Heritage Conservation Center of Beijing Guowenyan CO.,LTD, Tsinghua University-National Heritage Center (THU-NHC)
134	T1/ RADURA	Stefano Boeri	Stefano Boeri
142	T2/ 2017 The EU Mies Award and 2016'Fear of Columns'	LI Xiangning, GAO Changjun	Mies van der Rohe Foundation
148	T3/ Architecture's 'Political Compass': A Taxonomy of Emerging Architecture	Alejandro Zaera-Polo, G.Fernadenz	Alejandro Zaera-Polo, G.Fernadenz
150	T4/ The Mysteries of Columbus Cristobal	Cristobal Gabarron Foundation	Cristobal Gabarron Foundation
152	T5/ Cohesion	FANG Zhenning	FANG Zhenning
156	T6/ LOST IN A SHUFFLE	FANG Zhenning	Lawrence Weiner
162	T7/ Rolling Wind	FANG Zhenning	WANG Mai
164	T8/ Chinese Freestyle Scholars' Sculpture Park	FANG Zhenning	WU Weishan FANG Zhenning
170	T9/ Introspective Cavity (exterior)	GUO Xiaoyan	YIN Xiuzhen
174	T10/ Design Bench	FANG Zhenning	JIN Lie
176	T11/ Rhythm of Wind	LI Xiangning	SHENG Shanshan
178	T12/ Cang Sheng · Pin	LI Xiangning	SU Dan, WANG Ning, ZHANG Jian
180	T13/ 2340 holesThe	LI Xiangning	YU Xingze

	Exhibition Title	Curator	Exhibitors
184	H1/ Echo from Society: Architecture and Contemporary Challenges out of Established Agendas	Placido Gonzalez Martinez, TIAN Weijia	Francis Kéré, Jorge Raedó, Andrés Jaque, WANG Zigeng, Ecosistema Urbano, Rural Urban Framework, Instituto Pedra
190	H2/ Memory Project of the Nanjing Yangtze River Bridge	LU Andong	LanD Studio
194	H3/ Setouchi Triennale	Fram Kitagawa	LIN Shunlong
198	H4/ Connection: Space Movement	GUO Xiaoyan	LIU Wei, SHEN Yuan, QIU Zhijie, XU Zhen, HUANG Yongping, HE An, WANG Yuyang, Olafur Eliasson, LIU Jianhua, CHENG Ran, Hans op de Beeck, Dominique Gonzalez-Foerster, Fabien Giraud
214	H5/ The Frontier of Space	FANG Zhenning	WANG Yun
218	H5/ You are so tender, Fantasy Speed	FANG Zhenning	HUANG Ying
222	H5/ Malevich Finders	FANG Zhenning	Yung Ho Zhang
226	H5/ Cosmos	FANG Zhenning	LI lei
230	H5/ Madame Butterfly	FANG Zhenning	Lyu Yue (Aluna)
234	H5/ Future Metropolitan of AI Era	FANG Zhenning	HE Wanyu/XKool
238	H6/ INNER SPACE	FANG Zhenning	SHEN Wei, MA Yansong
246	I1/ Bridge	LI Xiangning, YAo Weiwei, XIE Yuqing	Atelier Deshaus, Wu Zhi Qiao (Bridge to China), Charitable Foundation, Atelier FCJZ, Heatherwick studio, James Corner Field Operations, Universidad Europea de Madrid
258	I2/ Interconnected in Production	Fab-Union	Fab-Union/YeArch Studio
262	I3/ Hyperloop TT Experience	LI Xiangning, DENG Yuanye	Hyperloop Transportation Technologies
266	I4/ Digital Fabrication: Digital Metal	Philip F. Yuan	Mania Aghaei Meibodi, Benjamin Dillenburger/ETH Zurich Digital Building Technologies
268	I4/ Digital Fabrication: Shells with Thin Sheet Materials	Philip F. Yuan	WANG Xiang/Fab-Union
270	I4/ Digital Fabrication: Robotic Timber Construction	Philip F. Yuan	Philip F. Yuan, CHAI Hua, MENG Gang/Fab-Union
276	S1/ Shanghai Practice	ZHI Wenjun, DAI Chun	Shanghai Municipal Bureau of Planning and Land Resources/Shanghai Tongji Urban Planning & Design Institute
281	S2/ Urban Humble-Space Regeneration	DAI Chun, YU Ting	Let's talk Academic Forum, WANG Yan, LIU Yichun, Shanghai Urban Planning & Design Research Institute, YU Ting, LI Yanbo, Atelier Archmixing, TONG Ming, JWDA, XI Wenlei, Fablab O
286	S3/ Shanghai Urban-Space Research	DAI Chun, FENG Lu	WANG Zhuoer, CAI Yongjie, ZHANG Jiajing, TAN Zheng/Tongji University Press
289	S4/ Study of New Villages	FENG Lu	Wuyang Architecture, Atelier Z+, Atelier Archmixing, Department of Architecture of Shanghai Jiaotong University, Atelier YeAS
293	S5/ Waterfront Connection	DAI Chun	Shanghai Municipal Bureau of Planning and Land Resources/Yangpu District, Pudong New Area, Hongkou District, Huangpu District, Xuhui District
300	S6/ Culture Enlightens the City	ZHI Wenjun, DAI Chun	Pudong New Area Planning and Land Authority; Shanghai Municipal Ad of Culture, Radio, Film and TV; Schmidt Hammer Lassen Architects; David Chipperfield Architects; Rurban Studio, Tongji Architectural Design (Group) Co., Ltd.; Atelier Li Xinggang, China Architecture Design & Research Group Co., Ltd.; Snøhetta; O'Donnell + Tuomey Architects; ATELIERS JEAN NOUVEL; Scenic Architecture Office; Original Design, Tongji Architectural Design (Group) Co., Ltd.; MVRDV
308	S7/ Renew Shanghai	DAI Chun	Shanghai East Bank Investment (Group) Co., Ltd.,

SUSAS 学院

SUSAS 学院是"上海城市空间艺术季"重点培育并打造的教育品牌。在活动期间,SUSAS 学院整合了各方资源,策划一系列课程/活动,搭建公众参与平台,发挥展览教育作用,一方面为专业人士提供一个交流的平台,另一方面希望能够感染和激发市民更多地了解关于城市、建筑、公共空间与公共艺术的相关知识,带动市民加入到对城市的了解、规划和更新当中,为在上海生活工作的市民塑造对城市的认同感和归属感。SUSAS 学院不是传统意义上的有物理空间的教育场所,而是有关城市、公共空间及公共艺术的面向大众的开放的教育资源,是协作学习及互动的社区、平台。

本届艺术季 SUSAS 学院活动总计 56 场。其中,学术讲座类 26 场,包括主策展人讲座、参展人讲座、社区课堂和论坛沙龙等;儿童教育类活动 23 场,包括小小导览员、小小观察员、儿童绘画和手工活动等。除此之外,艺术季主展场还有国际艺术节演出、未来之城中学生竞赛等 7 场公众活动。

SUSAS COLLEGE

SUSAS College is an educational program to which Shanghai Urban Space Art Season (SUSAS) devotes its efforts and assistance. During the event, the college gathers resources of parties together, organizing a lecture series and events, and establishes a public participatory platform for exhibition and education. It offers a forum for professionals, and also hopes to lead the public to further study about the city, its buildings, the public spaces and public art, engaging them to learn, plan and renovate the city, and furthermore shape a sense of belonging and appreciation among those working and living in Shanghai. Instead of a physical educational facility in the traditional sense, the college is a publicly accessible library that collects materials concerning cities, public spaces and public art, as well as an interactive, collaborative community platform.

Fifty-six SUSAS College events are held during SUSAS 2017, including 26 lectures (talks of chief curators and contributors, community workshops and forums) and 23 educational activities for children (Young Tour Guide, Young Observers, drawing and handcrafting), among others. Apart from those of College, seven more public events, including international entertainment show and "City of Future" competition for middle school students, are held in the site of main exhibition.

2017/10/21/ 简单水岸：超级水岸景观工程的理想与现实 / 主策展人李翔宁工作团队 / 克里斯蒂安·多布里克、邓圆也
Simple Waterfront: The vision and reality of the mega waterfront projects / Work team of Chief curator LI Xiangning / Christian Dobrick, DENG Yuanye

2017/10/21/ 读城 / Let's talk 论坛 / 庄慎、席子、俞挺、陈胤希、谢震纬
Read City / Let's Talk / ZHUANG Shen, XI Zi, YU Ting, CHEN Yinxi, XIE Zhenwei

2017/10/28/ 中国上海国际艺术节演出 / 上海国际艺术节组委会
Performance of China Shanghai International Arts Festival / The Organizing Committee of China Shanghai International Arts Festival

2017/10/29/ 小小少年城市导赏员培训课 / 上海彩虹青少年发展中心
Training Session for Young Tour Guides / Shanghai Rainbow Youth Development Center

2017/10/31/ 水岸新生 / Let's talk 论坛 / 章明、王卓尔、谭峥、田唯佳
Waterfront newborn / Let's talk / ZHANG Ming, WANG Zhuoer, TAN Zheng, TIAN Weijia

2017/11/ 小小少年城市导赏员展览导览服务，小小观察员（6场）/ 上海彩虹青少年发展中心
Exhibition Practice Session for Young Tour Guides; Young Observors (six sessions) / Shanghai Rainbow Youth Development Center

2017/11/ 趣城课堂 "小小规划师" 手作活动（6场）/ 上海城市公共空间设计促进中心 / 同济大学志愿者
Young Planners in Fascinating City (six handcraft sessions) / Shanghai Design & Promotion center for urban public space / Tongji University volunteers

2017/11/01/ 数字建造 连接未来Ⅰ/ I2 物联生产，I4 数字建造分策展团队 / 袁烽、本杰明·迪伦伯格、马尼亚·阿格海伊·梅伯蒂、王祥
Digital fabrication Future Connection / Curators team of I2 Interconnected in Production and Curators team of I4 Digital Fabrication: Digital Metal / Philip F·Yuan, Benjamin Dillenburger, Aghaei Meibodi Mania, WANG Xiang

2017/11/04/ 临港滴水湖景观七桥：21世纪的建筑、工程和景观设计 / 白德龙
Green Belt 7 Bridges in Lingang, Architecture, Engineering and Landscape in the XXI Century / Pedro Pablo Arroyo Alba

2017/11/04/ 南京长江大桥的物质、记忆及其当代重连接 / 南京大学建筑与城市规划学院，LanD 工作室 / 鲁安东
Nanjing Yangtze River Bridge: Matter, Memory and their Contemporary Re-connection / College of Architecture and Urban Planning, Nanjing University, LanD studio / LU Andong

2017/11/11/ 如何成为国际策展人？/ 主策展人方振宁
How to become an international curator / FANG Zhenning

2017/11/12/ 浦江两岸十二景：儿童插画长卷墙 / 那行 / 黄瑶瑶
Children's Creative Long Scroll of Illustration / Zero Degree Play Ground / HUANG Yaoyao

2017/11/12/ 邬达克建筑 / 上海城市规划展示馆 / 刘杰、杨蕾
Architecture of Hudec / Shanghai Urban Planning Exhibition Hall / LIU Jie, YANG Lei

2017/11/17/ 天山西路城市慢行 / 长宁实践案例展策展团队 / 张佳晶、钱欣、王卓尔、庄令晔、马佳雯
City non-motorized system of Tianshanxi Road / Curatorial team of Changning Site Project / ZHANG Jiajing, QIAN Xin, WANG Zhuoer, ZHUANG Lingye, MA Jiawen

2017/11/18/ 我的上海地图 / H1 建筑与社会展位策展工作团队 / 艾特利尔·伍兹 / 豪尔赫·雷多
Shanghai Map / Curators team of H1 The Echo from Society, Atelier Wiz / Jorge Raedo

2017/11/18/ 手工传习所《衍纸贺卡》/ 上海城市规划展示馆 / 穗穗
Handcraft Classroom: Paper Quilling Cards / Shanghai Urban Planning Exhibition Hall / SUI sui

2017/11/19/ 我的上海地图 / H1 建筑与社会展位策展工作团队 / 艾特利尔·伍兹 / 豪尔赫·雷多
Shanghai Map / Curators team of H1 The Echo from Society, Atelier Wiz / Jorge Raedo

2017/11/19/ 海派手工技艺传习所：雕刻时光城市橡皮章 / 上海城市规划展示馆 / 乐了
Shanghai Urban Planning Exhibition Hall / LE le

2017/11/25/ 微空间复兴 / Let's Talk 论坛 / 俞挺、张海翱、薄宏涛、傅志强、戴春
Micro Space Renewal / Let's Talk / YU Ting, ZHANG Haiao, BO Hongtao, FU Zhiqiang, DAI Chun

2017/11/25/ 石库门 / 上海城市规划展示馆 / 刘杰、薛靓
Shikumen / Shanghai Urban Planning Exhibition Hall / LIU Jie, XUE Liang

2017/11/26/ Garage 莫斯科当代艺术博物馆：艺术与建筑的十年 / 主策展人方振宁工作团队 / 白雪
Garage Museum of Contemporary Art / Ten years of Art and Architecture / Chief curator FANG Zhenning's work team / Snejana Krasteva

2017/11/28/ 意大利传统工艺复兴 / Let's Talk 论坛 / 李伟、胥一波、西尔维奥·费斯塔里、莫罗帕·特里尼
Regeneration of Italian triditional techniques / Let's Talk / LI Wei, XU Yibo, Silvio Festari, Mauro Patrini

2017/12/02/ 未来之城大赛（少儿活动）/ 未来之城上海地区组委会
"City of Future" Competition for Middle School Students / Organizing committee of "City of Future" Competition Shanghai

2017/12/02/ PLACES and SPACES 系列：筒仓生日会 / 倾唯文化传播 / 倪旻卿
A Series of PLACES and SPACES: Birthday party of silo / NI Minqing

2017/12/03/ 社区微更新：设计师来帮你们出主意 / 上海城市公共空间设计促进中心
Community Micro-Regeneration: Designers are Here to Help / Shang Urban Pubilc Space Design and Promotion Center

2017/12/03/ 拉斯维加斯、电影和运动的凝视 / 主策展人李翔宁工作团队 / 马提诺·斯泰瑞利、王骏阳
Las Vegas, Film, and the Mobilized Gaze / Work team of Chief curator LI Xiangning / Martino Stierli, WANG Junyang

2017/12/07/ 智慧空间艺术沙龙 / 上海市信息服务业行业协会
Intelligent Space Art Salon / Shanghai Information Services Association

2017/12/08/ 社区，让生活更美好：浦东缤纷社区实践案例展 / 浦东新区实践案例展展览团队 / 薛英平、奚文沁、于海、杨贵庆、王玺昌、倪倩
Better Community, Better Life: Site Project of Pudong Colorful Community / Curatorial team of Pudong New District Site Project / XUE Yingping, XI Wenqin, YU Hai, YANG Guiqing, WANG Xichang, NI Qian

2017/12/09/ 文化点亮城市 设计引领未来：浦东新区文化设施设计详解 / Let's Talk 论坛 / 陈蓉、罗伯特·格林伍德、陈晨、祝晓峰、陈立缤、李立、冯子鹏、米歇尔·欧斯莱
Culture Enlightens City, Design Leads Future: A Detailed Introduction to Designs of Cultural and Entertainment Facilities in Pudong New Areas / Let's Talk / CHEN Rong, Robert Freenwood, CHEN Chen, ZHU Xiaofeng, CHEN Libin, LI Li, FENG Zipeng, Michelle Osley

2017/12/10/ 共享创意筒仓：儿童插画改造 / 那行 / 黄瑶瑶、庄慎
Kids's Creative Design on Silos / Zero Degree Play Ground / HUANG Yaoyao, ZHUANG Shen

2017/12/10/ PLACES and SPACES 系列：城市，连接你我他

2017/10/21/ 深沪互联：专业媒体看深双 / Let's Talk 论坛 / 斯坦法诺·博埃里、方振宁、支文军、李立、杨猷锋、李晓峰、于冰、龚彦、向玲、张涵、王舒
UABB@SUSAS Architecture&Art Media@UABB FORUM / Let's Talk / FANG Zhenning, ZHI Wenjun, LI Li, YANG Zixing, LI Xiaofeng, YU Bing, GONG Yan, XIANG Ling, ZHANG Han, WANG Shu

2017/12/15/ 未来城市 / 主策展人斯坦法诺·博埃里
Future city / Stefano Boeri

2017/12/17/ 共享创意筒仓：儿童插画改造 / 那行 / 黄瑶瑶、庄慎
Kids's Creative Design on Silos / Zero Degree Play Ground / HUANG Yaoyao, ZHUANG Shen

2017/12/20/ 日常融入：公众与滨水空间互动 / H+A 华建筑 / 杨智勇、程愚、章明、黄斌、张淑萍、曹嘉明、沈立东、邢同和、李武英
Immersion into Daily Life: Interaction between People and Waterfront / H+A Architecture / YANG Zhiyong, CHENG Yu, ZHANG Ming, HUANG Bin, ZHANG Shuping, CAO Jiaming, SHEN Lidong, XING Tonghe, LI Wuying

2017/12/23/ 自由的建筑设计 / 主策展人方振宁工作团队 / 王昀、方振宁
Architectural Design towards freedom / Work team of Chief curator FANG Zhenning / WANG Yun, FANG Zhenning

2017/12/24/ 小小少年城市导赏员戏剧表演与颁奖 / 上海彩虹青少年发展中心
Performance and Award Session for Young Tour Guides / Shanghai Rainbow Youth Development Center

2017/12/28/ 艺术与户外媒体 / 渡爱联合展展览团队 / 刘颖彤、朱赢椿、金江波
Outdoor Media / Curatorial team of Joint exbition of "Ferry love" / LIU Yingtong, ZHU Yingchun, JIN Jiangbo

2018/01/06/ 建筑学如何介入城市：一种可能性尝试 / Let's Talk 论坛 / 范文兵、冯路、张斌、庄慎、曹永康、王卓尔
Architectural Intervention in Cities: An Experimental Approach / Let's Talk / FAN Wenbing, FENG Lu, ZHANG Bin, ZHUANG Shen, CAO Yongkang, WANG Zhuoer

2018/01/06/ 嗨翻筒仓 艺享滨江——新年定向挑战赛 / 上海城市公共空间设计促进中心
New Year Orienteering Challenge: Come to the Amusing Silo and Waterfront

2018/01/07/ 乡建思考：从乡村到野外 / 主策展人方振宁工作团队 / Let's talk 论坛 / 何崴、俞挺
Rural to wild : recent works / Work team of Chief curator FANG Zhenning, Let's Talk / HE Wei, YU Ting

2018/01/07/ 城市航拍日 / 复旦大学信息与传播研究中心，新华网上海频道，上海城市公共空间设计促进中心 / 孙玮、赵宝静、叶森、黄旦、周海晏、梁鸿儒、单颖文、翟轶羿、王侠、钱进、李杰、潘霁、杨敏、罗沛鹏、陆晔
Shanghai Aerial Photography Day / Center for Information and Communication Studies, Fudan University, sh.xinhuanet.com, Shang Urban Pubilc Space Design and Promotion Center / SUN Wei, ZHAO Baojing, YE Sen, HUANG Dan, ZHOU Haiyan, LIANG Hongru, SHAN Yingwen, ZHAI Yiyi, WANG Xia, QIAN Jin, LI Jie, PAN Ji, YANG Min, LUO Peipeng, LU Ye

2018/01/13/ 漫步两岸 乐享浦江
Walk and Enjoy Huangpu Riverside

2018/01/14/ 数字建造 连接未来Ⅱ I2 物联生产，I4 数字建造分策展团队 / 袁烽、吴迪、张周捷、Fablab
Digital fabrication Future Connection / Curators of I2 Interconnected in Production and Curators of I4 Digital Fabrication: Digital Metal / Philip F·Yuan, WU Di, ZHANG Zhoujie, Fablab

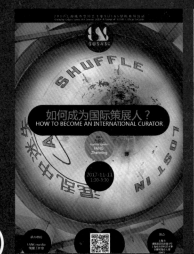

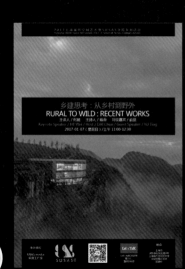

策展团队

主策展人 / CHIEF CURATORS

斯坦法诺·博埃里 / 意大利
著名建筑师、评论家、教育家及策展人。目前担任意大利米兰理工大学城市设计教授，曾执教于美国哈佛大学设计学院、哥伦比亚大学、麻省理工学院等建筑院校。

Stefano Boeri
A well-known architect, critic, educator and curator, Stefano is now a professor of Urban Design at Politecnico di Milano. He has taught courses and given lectures at Harvard University Graduate School of Design, Columbia University, MIT and other architectural institutes.

李翔宁 / 中国
同济大学建筑与城市规划学院副院长、教授、博士生导师。国际建筑评论家委员会委员，曾担任2013深圳双年展策展人和学术总监、上海西岸双年展策展人，欧盟当代建筑奖密斯·凡·德·罗奖、西班牙国际建筑大奖等国际建筑奖项评委。

LI Xiangning
Vice Dean, professor and doctoral supervisor of the School of Architecture and Urban Planning at Tongji University. He is a member of the International Committee of architectural critics, and has served at the 2013 Shenzhen Biennale and Academic Director, the Westbund Shanghai Biennale curator, as well as jury member of Spain International Architecture Award, Mies van der Rohe award the European Union Prize for contemporary architecture and other International Architecture Awards.

方振宁 / 中国
独立策展人、教授、艺术批评家，艺术和建筑领域的独立学者。2012年第13届威尼斯建筑双年展中国国家馆策展人，2014年威尼斯建筑双年展参展人、中国美术家协会策展委员会委员。

FANG Zhenning
Independent curator, professor, art critic, and independent scholar in the fields of art and architecture. Curator of China Pavilion in the 13th Venice Architecture Biennale 2012, exhibitor of the 14th Venice Architecture Biennale 2014, and a member of the curatorial committee of the China Artists Association.

联合策展人 / JOINT CURATORS

郭晓彦 — GUO Xiaoyan
冯路 — FENG Lu
支文军 — ZHI Wenjun
戴春 — DAI Chun
托马索·萨基 — Tommaso Sacchi

策展团队 / CURATORIAL TEAM

时代建筑 — *Time+Architecture*
Let's Talk 学术论坛 — Let's Talk Forum

助理策展人 / ASSISTANT CURATORS

姚微微 — YAO Weiwei
田唯佳 — TIAN Weijia
莫万莉 — MO Wanli
毛甜甜 — MAO Tiantian
王澍 — WANG Shu
胥一波 — XU Yibo
高长军 — GAO Changjun

建筑改造前 Before building renovation

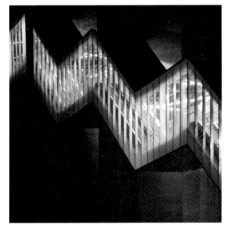

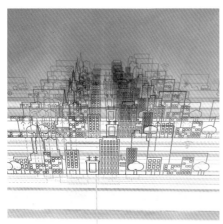

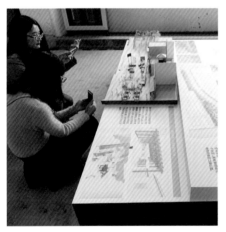

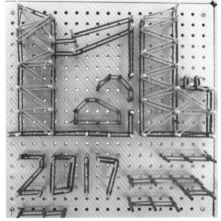

图片版权与摄影者索引
COPY RIGHTS AND PHOTOGRAPHY CREDITS

摄影照片提供（按首字母顺序排序）
PHOTOGRAPHY CREDITS (Alphabetical order by name)

方振宁 /FANG Zhenning
P6-7，P10-11，P12-13,P14-15，P16-17,P18，P31，P34-35，
P58-59，P70-71，P72，P76-77，P80-81，P82，P83，P84-85，P86，
P88-89，P96-97，P98，P106-107，P108，P109，P110-111，P113，
P114-115，P117，P118-119，P120，P121，P124-125，P128-129，
P131，P134-135，P136-137，P139，P140,P141,P142-143，
P146-147，P149，P150-151，P152-153，P154，P155，P156-157，
P159，P162，P163，P164-165，P166，P167，P168，P169，
P170-171，P172，P173，P174-175，P176-177，P180-181，
P184-185，P190-191，P192，P193，P194-195，P196，P198-199，
P200，P201，P202，P203，P204，P205，P206，P207，P208，
P209，P211，P212，P213，P214-215，P217，P218-219，
P222-223，P224，P225，P226-227，P228，P229，P230-231，
P232，P234-235，P236，P240，P246-247，P249，P258-259，
P261，P262-263，P266-267，P270-271，P276，P277，P281，
P282，P286，P287，P289，P293，P300，P301，P308，P317，
P324，P326，P327，P338-339，P340-341，P344-345

高长军 /GAO Changjun
P210

金选民 /JIN Xuanmin
P178-179，P233，P268-269

田方方 /TIAN Fangfang
封面/Cover，P4-5，P54-55，P250，P342-343，P346-347

薛长命 /XUE Changming
P2-3

严帅帅 /YAN Shuaishuai
P8-9

万科集团 /Vanke group
P348-349

花絮图片提供：上海城市公共空间设计促进中心 / 方振宁 / 金选民
Tidbits Photo Provider : Shanghai Design & Promotion Center for Urban Public Space / FANG Zhenning / JIN Xuanmin

致谢 ACKNOWLEDGEMENTS

2017第二届上海城市空间艺术季主展览经过一年多的筹备,于2017年10月15日开幕,2018年1月15日闭幕。来自世界各地的200多位规划师、建筑师和艺术家围绕主题"连接 thisCONNECTION: 共享未来的公共空间",充分展示了他们各自的作品和研究,为城市公共空间中的"断裂"提供了多样的、有效的"连接"可行性。主展场在三个月内共接待12万多的参观人次,成功举办了各类SUSAS活动和其他公众活动,活动类型丰富,覆盖人群广,市民参与度高。空间艺术季活动感染和激发了各个年龄层次的市民对城市、建筑、公共空间与公共艺术的认知与兴趣,传递了建成卓越的全球城市,令人向往的创新之城、人文之城、生态之城的城市发展愿景。

与上一届的主展场徐汇西岸艺术中心不同,本届主展场选择在浦东新区停产多年的民生码头8万吨筒仓及周边开放空间,旨在通过艺术介入方式,激发地区活力。在有限的时间内,展场需要完成一期改造,策展方案需要实施,展览需要按期开幕,为此,展场改造建筑师,施工建设单位,展场运营方、策展团队和执行团队精诚团结、金石为开,共同完成了这项几乎不可能完成的任务。

在此,特别鸣谢以下单位和团体:
提供场地的上海地产(集团)有限公司;为展场提供相关业务指导的浦东新区宣传部(文广局)、浦东新区公安局、浦东新区绿化市容局、上海陆家嘴金融贸易区管理委员会、洋泾街道等;展场建设单位上海建工一建集团有限公司,展场运营单位上海木宁芙文化发展有限公司;为活动宣传提供广告阵地的上海申通地铁集团有限公司、郁金香广告传播(上海)有限公司、协助活动报道的解放日报、文汇报、新民晚报、上海电视台等传统媒体。

抚今追昔,感慨于心。上海城市空间艺术季在各方的鼎立支持下正逐步成为有影响力的活动和品牌。在未来的日子里,我们希望能够与各方继续携手并进,共同发展这一城市品牌活动,为弘扬上海城市文化,打造具有世界影响力的社会主义现代化国际大都市增光添彩!

The second Shanghai Urban Space Art Season, SUSAS 2017, was launched at 15 October after preparation of more than one year, and will end at 15 January, 2018. Centering on the theme, "thisCONNECTION: Sharing a Future Public Space", over 200 planners, architects and artists from all over the world have developed their results of creativity or research, offering a variety of plausible approaches to "connect" the "disconnected" points in urban public spaces. Besides the main exhibition has been visited by more than 120,000 people in the three months, various related public activities, SUSAS or not, are held that attract and engage citizens of considerable number and comprehensiveness.

The SUSAS events have infected and inspired citizens of all ages to learn and concern cities, buildings, public space and public art. At the same time, the appealing vision of Shanghai to be an outstanding global city featuring innovation, humanity and environmentalism is widely spread among all communities. In contrast to Art West Bund in Xuhui District, which hosted the first SUSAS, the long-abandoned 80,000-ton Silo and its adjacent open fields in Pudong New Area is chosen as the main exhibition site of SUSAS 2017, intended to energize the neighborhood with art intervention. Albeit the limited time span within which renovation, curation and exhibition on-site installation need to be done all together, the renovation architects, constructor, site operator, curators and execution team have completed this impossible mission with incredible competence and cooperativeness.

Hereby extends its gratitude to:
Site provider: Shanghai Land (Group) Co., Ltd
Service facilitators: Publicity Department of Pudong New Area (Administration of Culture, Radio, Film & TV), Pudong Branch of Shanghai Public Security, Pudong New Area Afforestation & City Appearance and Environmental Sanitation Administration, Lujiazui Financial and Trade Zone Management Committee, Yangbang Sub-district Administration, etc.
Constructor: Shanghai Construction No.1 (Group) Co., Ltd.
Site Operator: Shanghai Muningfu Cultural Development Co., Ltd.
Advertisers: Shanghai Shentong Metro Co., Ltd, Tulip Media, China Huaneng Group, etc.
Press Partners: Jiefang Daily, Shanghai Wenhui Daily, Xinmin Evening News, Shanghai Media Group and other traditional media, and Shanghai Post (Weibo account).

Memories and sentiments are many as we reflect on the process in which SUSAS rises as an influential art season with the support from various parties. In the future, we expect to further develop the shining name card of Shanghai with our partners, and add glamor of culture to this modern, international and socialist metropolitan that aims at global significance.

2017上海城市空间艺术季
导 览 地 图

实践案例展 静安区

"双重曝光·旧厂新生"
新业坊　汶水路210号新业坊

实践案例展 长宁区

新泾·新境
长宁区民俗文化中心　北渔路95号

实践案例展 徐汇区

"为风貌而设计"　黑石公寓　永嘉新村

联合展 青浦区

实践案例展 徐汇区

融"汇"贯通
T20大厦　天钥桥路20号

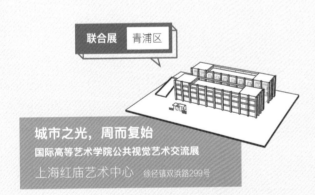

城市之光，周而复始
国际高等艺术学院公共视觉艺术交流展
上海红庙艺术中心　徐泾镇双浜路299号

"双向辐射"城市更新在
西岸公共开放沿线
星美术馆、油罐艺术公园、跑道公园、
西岸艺术中心周边

上海城市空间艺术季执行团队

上海市规划和国土资源管理局风貌处：侯斌超、林磊、应孔晋

上海市文化广播影视管理局艺术处：沈捷、沈雪江

上海城市公共空间设计促进中心：赵宝静、王明颖、陈敏、胡颖蓓、王夏娴、周敏、马宏、陈成、薛娱沁、赵起超、严帅帅、章舒雯

浦东新区规划和土地管理局：关也彤、刘伟、汤明华、魏文、赵波

上海东岸投资（集团）有限公司：黄磊、程世红、韩莹赟、王芳、屈伯禹、曾雨佳、陈亮、孙政

官方网站：http://www.susas.com.cn

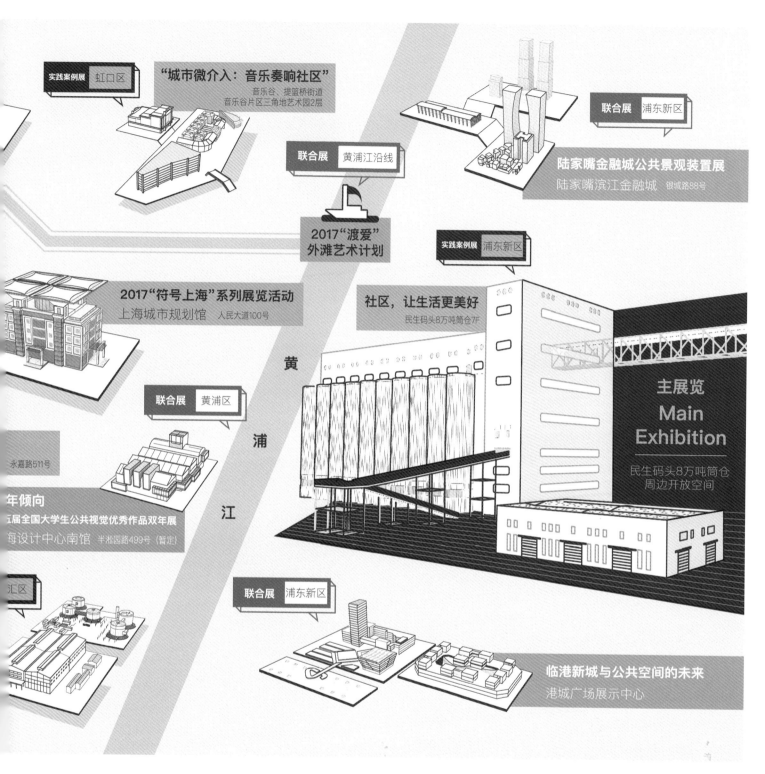

SUSAS Execution Team

Landscape Division, Shanghai Planning and Land Resource Administration Bureau: HOU Binchao, LIN Lei, YING Kongjin

Art Division, Shanghai Municipal Administration of Culture, Radio, Film & TV: SHEN Jie, SHEN Xuejiang

Shanghai Design & Promotion Center for Urban Public Space: ZHAO Baojing, WANG Mingying, CHEN Min, HU Yingbei, WANG Xiaxian, ZHOU Min, MA Hong, CHEN Cheng, XUE Yuqin, ZHAO Qichao, YAN Shuaishuai, ZHANG Shuwen

Planning and Land Resource Administration Bureau of Pudong New District, Shanghai: GUAN Yetong, LIU Wei, TANG Minghua, WEI wen, WEI Wen, ZHAO Bo

Shanghai East Bund Investment (Group) CO., LTD: HUANG Lei, CHENG Shihong, HAN yingyun, HAN Yingyun, WANG Fang, ZENG Yujia, CHEN Liang, SUN Zheng.

图书在版编目（CIP）数据

连接：共享未来的公共空间：2017 上海城市空间艺术季主展览 /
上海城市空间艺术季展览画册编委会编．
-- 上海：同济大学出版社，2018.6
ISBN 978-7-5608-7581-1

Ⅰ.①连... Ⅱ.①上... Ⅲ.①城市规划 - 空间规划 -
上海 - 画册 Ⅳ.① TU984.251-64

中国版本图书馆 CIP 数据核字 (2017) 第 319486 号

连接：共享未来的公共空间
2017 上海城市空间艺术季主展览

上海城市空间艺术季展览画册编委会 编

执行主编：方振宁
　　　　　李翔宁
　　　　　赵宝静

出 版 人：华春荣
策　　划：上海城市公共空间设计促进中心　方振宁

项目统筹：马　宏，胡颖蓓
责任编辑：秦　蕾，李　争
责任校对：徐春莲
翻译审核：王夏娴
封面设计：方媒体
装帧设计：方媒体（艺术总监：方振宁；设计执行：毛甜甜）

版　　次：2018 年 6 月第 1 版
印　　次：2018 年 6 月第 1 次印刷
印　　刷：北京翔利印刷有限公司
开　　本：889mm×1194mm 1/16
印　　张：22
字　　数：704 000
书　　号：ISBN 978-7-5608-7581-1
定　　价：258.00 元
出版发行：同济大学出版社
光明城联系方式：info@luminocity.cn
地　　址：上海市四平路 1239 号
邮政编码：200092

网　　址：http://www.tongjipress.com.cn

本书若有印装问题，请向本社发行部调换
版权所有 侵权必究

thisCONNECTION : SHARING A FUTURE PUBLIC SPACE
2017 SHANGHAI URBAN SPACE ART SEASON MAIN EXHIBITION

Edited by: Shanghai Urban Space Art Season Album Editorial Committee

Executive Editors:
FANG Zhenning
LI Xiangning
ZHAO Baojing

ISBN 978-7-5608-7581-1
Publisher: HUA Chunrong
Initiated by:
Shanghai Urban Public Space Design Promotion Center / FANG Zhenning

Project Coordinator: MA Hong / HU Yingbei
Editor: QIN Lei / LI Zheng
Proofreading: XU Chunlian
Translation Review: WANG Xiaxian
Graphic Design: FANGmedia
(Art Director: FANG Zhenning; Design Executive: MAO Tiantian)

Published in June 2018, by Tongji University Press,
1239, Siping Road, Shanghai, China, 200092.
www.tongjipress.com.cn
Contact us: info@luminocity.cn

All rights reserved